I. Strachan
11/4/73.

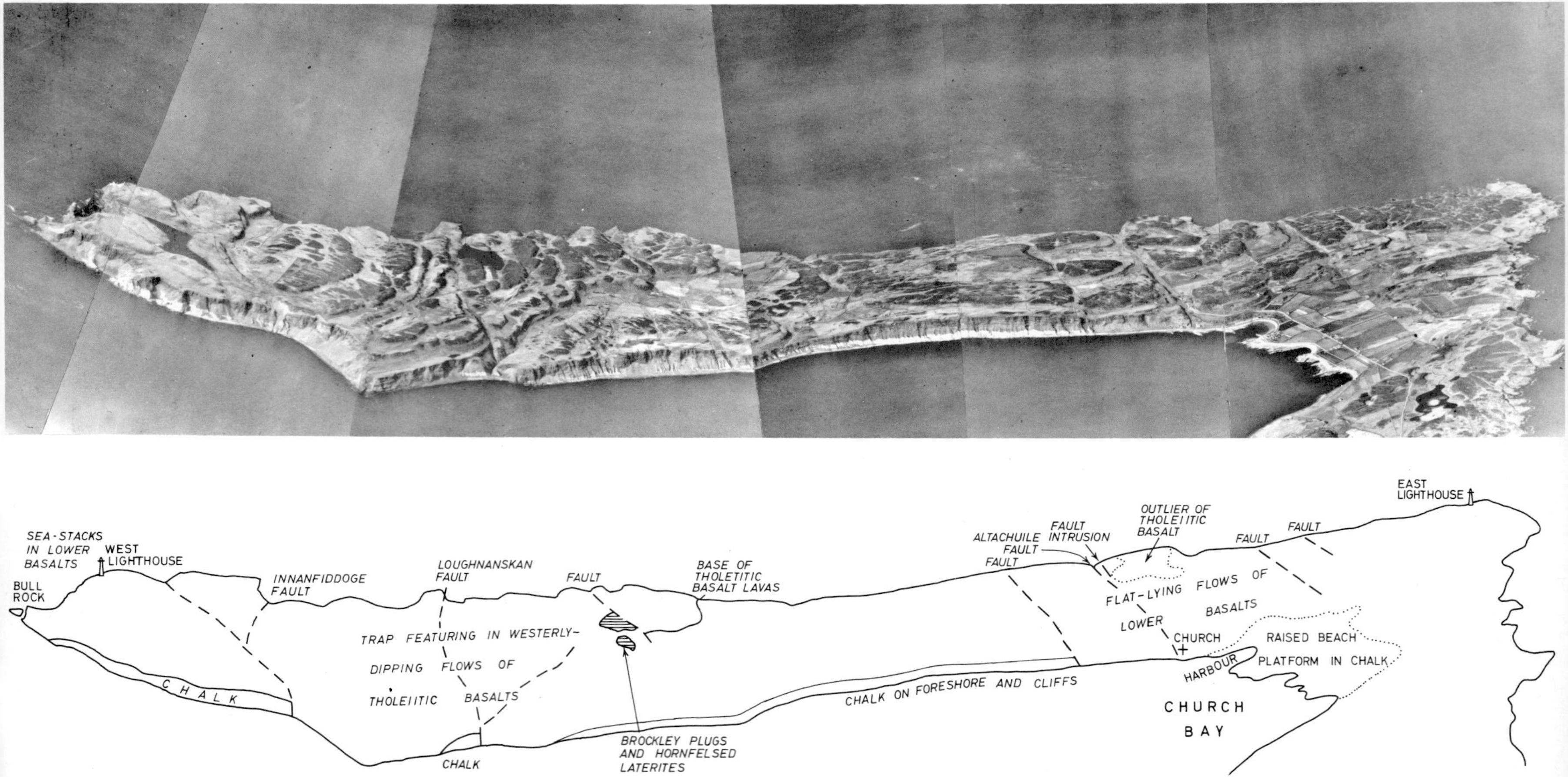

Oblique aerial photo-mosaic of Rathlin Island - RAF photographs.

MINISTRY OF COMMERCE
GEOLOGICAL SURVEY OF NORTHERN IRELAND

Regional Geology of

Northern Ireland

By H. E. Wilson, M.Sc.

BELFAST
HER MAJESTY'S STATIONERY OFFICE
1972

The Geological Survey of Northern Ireland is operated on behalf of the Ministry of Commerce by the Institute of Geological Sciences; a constituent body of the Natural Environment Research Council.

First published 1972

Foreword

This volume extends to Northern Ireland the series of handbooks on the Regional Geology of Great Britain which for 30 years have been of constant value to amateur and professional geologists as concise but informative reference books. Because of the extraordinary variety of geological formations in Ulster it has not been possible, within the format of the series, to describe every formation in the detail which might have been wished but it is hoped that the handbook will be of value, not only to geologists but to all who are interested in the natural history of Ulster.

Where opinions expressed or facts stated are not supported by published references it may be assumed that they are the results of work in progress by the Geological Survey. The sketch maps and diagrams are not intended as substitutes for published maps, a list of which is given at the back of the book. It is expected that a new map of Ulster on the scale of 1/250 000 will be published soon after this handbook.

In addition to the author other members of the staff of the Geological Survey of Northern Ireland have contributed to this publication. R. A. B. Bazley wrote much of the chapter on the Quaternary and contributed to that on the Metamorphic Rocks. I. B. Cameron provided material incorporated in the chapter on the Carboniferous; E. J. Cobbing on the Metamorphic Rocks and the Caledonides; T. P. Fletcher on the Cretaceous and P. I. Manning on the Permo-Trias and the Cretaceous.

C. J. Wood of the Institute of Geological Sciences has written the chapter on the Cretaceous and prepared the accompanying text-figure, and P. T. Warren contributed to the chapter on the Lower Palaeozoic.

Thanks are due to Dr. A. G. Smith and Dr. N. Stephens of Queen's University, Belfast, for generous assistance.

Most of the photographs are from the Geological Survey collection, identified by N.I. numbers.

Photographs by Dr. J. K. St. Joseph are reproduced by permission of the University of Cambridge Committee for Aerial Photography (Cambridge University Collection: copyright reserved).

K. C. DUNHAM
Director

Geological Survey of Northern Ireland,
20 College Gardens,
BELFAST, BT9 6BS.

Contents

Illustrations

Figures in Text

Tables

Plates

Numbers prefixed by N.I. refer to the Geological Survey photograph collection.

Photographs by J. K. St. Joseph are reproduced by the permission of the University of Cambridge Committee for Aerial Photography (Cambridge University Collection: copyright reserved).

I. Introduction

The area described here is the political unit of Northern Ireland and comprises six of the counties of the Province of Ulster—Antrim, Down, Londonderry, Tyrone, Armagh and Fermanagh. It excludes the adjacent counties of Donegal, Cavan and Monaghan, which are part of the Republic of Ireland, but some account of the geology of these areas will necessarily be included in the description of the southern and western parts of Fermanagh, Londonderry and Tyrone.

Within this region the geology is of a variety unsurpassed in any other district of similar size in the British Isles, including every geological system from the Moinian to the Quaternary, with the exception of the Cambrian, together with a wide variety of glacial and recent phenomena. Structurally, the area is to a great extent a south-westerly extension of Scotland, though the sharp topographical divisions made by the great faults which bound the Central Valley of Scotland are somewhat blurred in Ulster by the occurrence of Carboniferous, Mesozoic and Tertiary formations which overlap the Caledonian boundaries.

The foundations on which the region is built are the Dalradian and Moinian metamorphic rocks which appear in north-east Antrim, disappear under the Tertiary lava plateau, and reappear in the Sperrin Mountains and Donegal, still with the dominant Caledonian trend which moulds the Scottish Highlands. South of the line of the Lagan Valley the continuation of the Southern Uplands of Scotland is seen in the ancient landscape of Down and Armagh where Ordovician and Silurian rocks strike south-west into Monaghan and Cavan and on towards the west coast.

The Old Red Sandstone and Carboniferous rocks which once spread far to the north of the Highland Border in Ireland are now largely confined to the area between the faults with outliers to the north, though in the west the Carboniferous overrides all the earlier formations and extends far beyond the Highland Boundary Fault.

The dominant feature in the north-east is the wide Antrim plateau of Tertiary basalt lavas. This has acted as a protective carapace to the softer Cretaceous, Jurassic and Triassic rocks which now crop out round the tattered fringes of the plateau.

Over the whole of the region the relics of the Pleistocene glaciation profoundly affect the scenery and human geography. The whole area, save for the highest hills, is blanketed by glacial drifts which obscure the underlying rocks and control the pattern of settlement and agriculture. Ice from Scotland played a major part in the glaciation of the eastern part of the country, but in the west, Irish Ice from Donegal and possibly from centres in the Sperrins, was dominant. The advance and decay of these ice-sheets have left their mark in morainic deposits, the relics of glacial lakes, and the moulding of the landscape.

Regional Geography and Scenery

The region may be divided into several sub-units of markedly different aspect in which the topography and scenery reflect the underlying geological structure (Fig. 1, Plate 18).

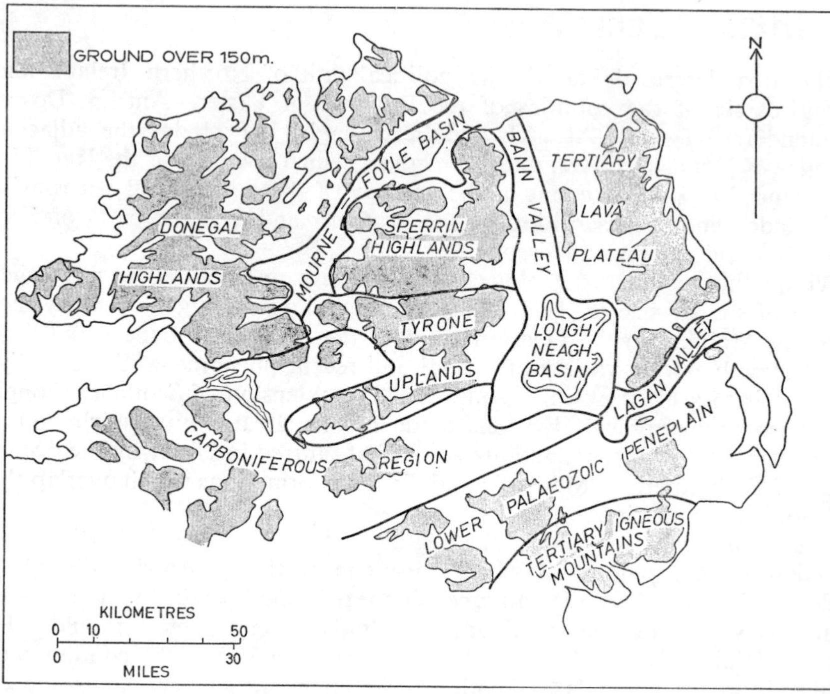

FIG. 1. *The natural regions of Ulster*

The Tertiary lava plateau of Antrim and east Londonderry is gently folded on a north–south axis to form the long valley of the Lower Bann. To west and east the plateau rises to heights of well over 300 m and on all sides except the south-west presents steep scarp faces to the outside world. Along the north and east Antrim coasts the basalts and underlying Chalk give the rugged coastal scenery for which the county is famous: to the west the basalt scarp towers over the plain of north Derry; and in the south the Lagan Valley, the centre of half the population of the province, is dominated by the basalt hills. Only round Lough Neagh does the lava country subside to scenic impotence.

The Lough Neagh basin is a tectonic feature and there is some indication that it has been an area of subsidence since Jurassic times. The Lough itself, the largest fresh-water lake in the British Isles, is scenically disappointing as it lies on a flat plain floored by Oligocene Lough Neagh Clays and down-faulted Tertiary basalt lavas, and only long-range views across to the hills on the west give any impression of its size. The basin extends along the valleys of the Blackwater and Upper Bann and eastwards into the Lagan Valley which is floored by soft Triassic rocks and overlooked by the lava scarp to the north and the uplands of Co. Down to the south.

South of the line of the Southern Uplands Fault which runs south–west from Belfast the northern parts of Down and Armagh is a low peneplain of Lower Palaeozoic rocks with little marked relief save for the exceptional development of swarms of boulder clay drumlins. The Caledonian Granite mass of Slieve Croob rises as an upland area in the centre of Down.

The southern part of Down and Armagh has the region of highest relief and most striking mountain scenery in the province, in the Tertiary granite mountains of the Mournes and the basic intrusive rocks of Slieve Gullion, with its encircling ring-dyke. This igneous complex is completed by the intrusive acid and basic rocks of the Carlingford peninsula in Co. Louth on the south side of Carlingford Lough. Carlingford Lough is a typical fjord and the area is one of exceptional scenic beauty.

West of the Tertiary lava plateau, and stretching north of the Highland Border Fault as far as the western limits of Donegal, is a wide area of Dalradian and Moinian schists with a few outliers of more recent rocks. In the west this region has a strong Caledonian trend, induced in part at least, by numerous north-east trending wrench faults, but east of the Foyle-Mourne-Strule basin the grain of the country is ill defined though the trend is still Caledonian. The scenery in these two sub-regions is markedly different. In Donegal a highly dissected terrain with rugged pinnacles and ridges gives way in Londonderry and Tyrone to a more gentle topography with relics of a high plateau forming rounded hills—the Sperrin Mountains.

South of the high ground of the Sperrin a wide area of moderate relief and poor soils on Old Red Sandstone falls away to the gentle lowlands of the Clogher Valley and the Erne Basin on the Carboniferous Limestone. To the south and west, outliers of Upper Viséan grit form hilly areas along the Eire border. West of Lough Erne in the foothills of Cuilcagh, striking scars of Carboniferous Limestone with swallow-holes and pots give a type of karst scenery unknown elsewhere in Ulster.

Within the confines of this small province, therefore, there is scenery to match the Highlands and Western Isles of Scotland, the Pennines, the Chalk cliffs of Yorkshire and the English Channel, and, on a smaller scale, drumlin swarms, unequalled in Europe, on the Lower Palaeozoic peneplain of Down and Armagh.

Soils and Agriculture

Though virtually the whole surface of the lowlands of Ulster is mantled by thick glacial drift which often stretches far up the hills, the soils reflect the character of the underlying rocks. This is especially so in the areas covered with boulder clay, where the drift is generally local in origin, but less so in the areas of glacial sands and gravels. The soils in the basaltic area are often reddish in colour though dark and brown soils are common. With good management they form good pasture but have a high capacity to fix soluble phosphates into unavailable forms. In the Lower Palaeozoic area the soils are very stony but form good agricultural land. In the schist areas, on the other hand, the extremely acid soil, unless well managed, gives poor returns.

In surprising contrast to Scotland, the Old Red Sandstone in Co. Tyrone gives rise to heavy waterlogged ground which is poor in plant-food minerals and is so clayey as to defy improvement. The soils on the Carboniferous Limestone are in general good but those on the other Carboniferous rocks—shales and sandstones—are poor.

The areas underlain by the Triassic sandstones in the Lagan Valley and south-west of Lough Neagh, and along the west side of the basalt scarp in

Londonderry offer the best arable land in the province but the marls tend to produce a heavy soil more suitable for pasture.

The cool and damp climate in Ulster, with rainfall evenly distributed throughout the year, results in intensive leaching from the soils of lime and magnesia and the production of waterlogged soils and peats. Over most of the province the emphasis is on mixed farming with an increasing tendency to pasture, and only in the east of the province is there much arable farming though the potato is widely grown. The extensive cultivation of flax, once the foundation of the linen industry, has virtually ceased.

History of Early Research

The clearly displayed geology of the Antrim Coast attracted early attention from a number of observers who wrote accounts of travels in the north-east in the late eighteenth and early nineteenth centuries, many spurred to visited the area by the controversy over the 'Portrush Rock'—hornfelsed fossiliferous Lias shales which were thought to resemble basalt. Whitehurst, Berger, Conybeare and Buckland, among others, gave good accounts of the basalts and the Chalk.

The first systematic examination of the whole province was due to Richard Griffith who himself, and later through his subordinates in the Valuation Office, covered much of the country and described it in his Reports on the Coalfields of Tyrone and Antrim (1829) and Report to the Railway Commissioners (1837).

With the commencement of the Ordnance Survey of Ireland in 1832 an enlightened director, Colby, proposed to conduct a simultaneous geological and ethnographical survey which would have resulted in a complete picture of the country. Financial stringency stopped this ambitious project after the production of only one Memoir—that on the parish of Templemore (Londonderry) in 1837—but Captain J. E. Portlock who was in charge of the Geological Branch of the Ordnance Survey was given permission to write up the work that he and his small staff had completed. The result was "*The Geology of the County of Londonderry and of parts of Tyrone and Fermanagh*" which is, by any standards, and particularly those of the time, a monumental work and which is still in many districts the most comprehensive account of the local geology.

In the latter half of the century the Geological Survey of Ireland covered the province with its series of excellent one-inch to the mile maps and generally rather inadequate memoirs, which were supplemented after the turn of the century by more detailed memoirs on the Londonderry and Belfast districts, the Iron Ores of Antrim and the Ballycastle Coalfield. Among the early surveyors, the work of DuNoyer, Hardman, Hull, Kilroe, Nolan and Symes was outstanding.

References

Many publications cover more than one of the topics discussed in the several chapters in this book. To avoid needless repetition the selected references quoted for each chapter are given in abbreviated form. The complete reference may be found in the Bibliography at the end of the book.

Berger 1816; Charlesworth 1953, 1963b; Cole 1912, 1924; Evans 1952; George 1960, 1967; Hull 1878; Lamplugh 1904; Portlock 1843; Whitehurst 1786. Wilkinson 1908;

A. *Boudinage, caused by tensional stress during metamorphism, in Dalradian quartz-schists. Portnadig, Torr Head.*

PLATE 1

B. *Folded Silurian greywackes and shales. Woburn House, Millisle.*

A. *Sole structures—loaded flute casts—on inverted Silurian greywacke. Near Galloways Burn, Donaghadee.*

PLATE 2

B. *Grooves and small load-casts on Ordovician greywacke. Craigavad, Co. Down.*

2. Metamorphic Rocks— The Crystalline Caledonides

Problem areas along the Highland Border

North of the Highland Border Fault the Sperrin schists are correlated on good evidence with the Dalradian rocks of Scotland, but there are some areas along the Highland Border zone in Ireland where the history of the rocks is as yet uncertain. These are:

1. The igneous and metamorphic complex of central Tyrone
2. The Lough Derg Psammites
3. The Migmatitic succession of the Ox Mountains
4. The gneisses of Co. Mayo.

Recent work in the Castlebar and Belmullet areas has suggested that certain gneisses and migmatites there may be of Lewisian age and it is inferred that some of the migmatitic and high-grade metamorphic rocks of the Highland Border may be of pre-Caledonian origin. On the contrary, however, a complex of high-grade metamorphic rocks, with associated migmatites and basic and acid magmatism, in Connemara, lying south of the Highland Boundary, is of undoubted Dalradian age.

Cobbing has suggested that the high grade of metamorphism in the metasediments of the Tyrone complex, with the associated basic and acid magmatism, indicated contemporaneity with the Connemara Schists and assigned them to the Lower Dalradian. He recognized, however, that the structural history of the Highland Border would not preclude the possibility that Moinian or Lewisian rocks could crop out south of the fault and represent a crystalline basement marginal to the Dalradian geosyncline which developed to the north-west. The Lewisian rocks of Inishtrahull, 9 km north of Malin Head, may represent the other flank of the geosyncline.

Only the Lough Derg and mid-Tyrone areas are within the district covered by this account.

The Lough Derg Psammites

A roughly triangular area of over 200 sq km between Lough Erne and Lough Derg is bounded on the south and west by overlying Carboniferous sediments, on the east by a major fault, and on the north by overthrust Dalradian rocks (Fig. 2). The rocks of this area are the Lough Derg Psammitic Group described by Anderson and are exposed in low crags of bare rock with scarp faces to the south. They consist largely of quartz-feldspar-mica-granulites in which the quartz is usually dominant, but beds of feldspathic granulite are not uncommon. The feldspars are usually microcline and oligoclase with rare orthoclase. Marked muscovite laminae a few centimetres apart give a pronounced flaggy appearance on weathering. Accessible outcrops occur along the road from Pettigo to Laghey, on the moorland north of the River Erne east of Ballyshannon, and north of Lough

Scolban [G 990 605]. Occasional beds of pelitic schist, from a few centimetres to a few metres thick, are present and these are strongly foliated, unlike the granulites. A band in which the pelitic schists are dominant can be traced from Ballyshannon to Lough Scolban.

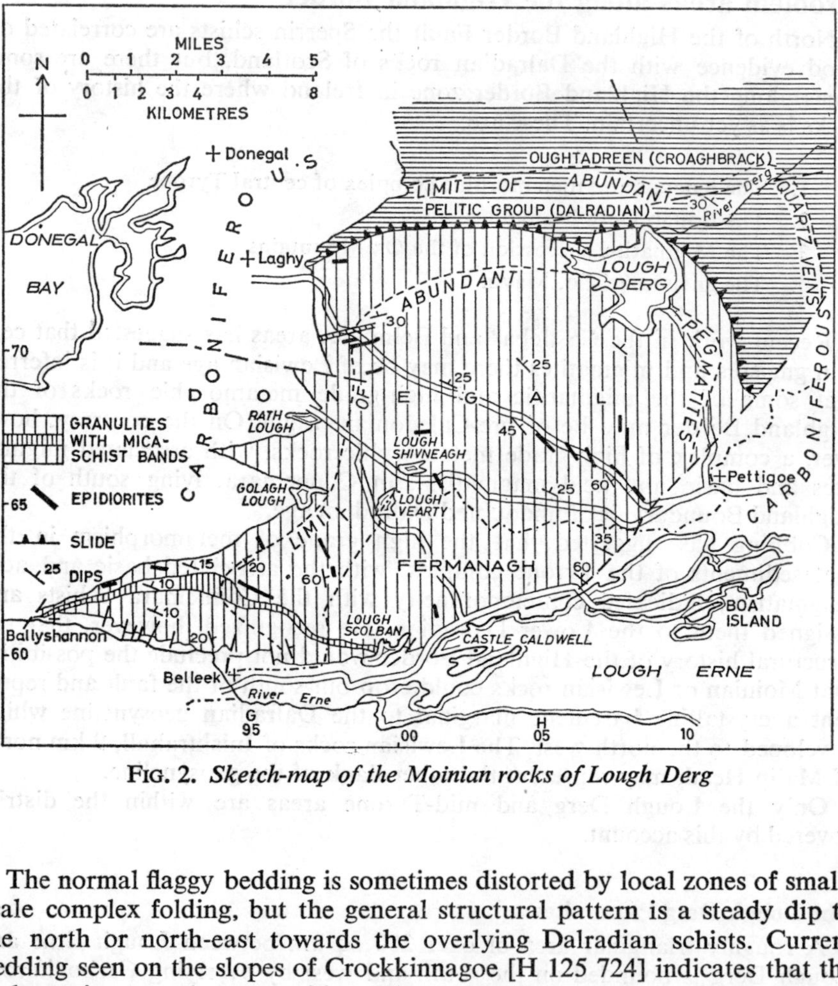

Fig. 2. *Sketch-map of the Moinian rocks of Lough Derg*

The normal flaggy bedding is sometimes distorted by local zones of small-scale complex folding, but the general structural pattern is a steady dip to the north or north-east towards the overlying Dalradian schists. Current bedding seen on the slopes of Crockkinnagoe [H 125 724] indicates that the beds are in normal superposition. The main structural feature is the Lough Derg Anticline, a broad fold with north–easterly axis plunging steeply north-east.

The granulites are penetrated by numerous intrusions of metamorphosed basic igneous rocks which form lenticular beds, generally only a few metres thick, which are probably the result of plastic deformation of sill-like intrusions in the original sediments. The rocks are dark massive epidiorites and amphibolites, rich in hornblende. There are occasional occurrences of garnet-rich amphibolites. Good exposures occur along the Pettigo–Laghey road.

The Moinian rocks are extensively veined by a series of microcline-oligoclase-quartz-mica-pegmatites and quartz stringers, injected during the Caledonian earth movements. The pegmatite bodies are very irregular and lenticular and are usually 0.3 to 2 m wide. Easily accessible examples are seen along the road from Pettigo to Laghey as at [H 070 677], near Lough Unshin [G 942 633] where they were worked for feldspar, and at Larkhill [H 013 633].

The Igneous and Metamorphic complex of Central Tyrone

The complex may be readily divided into three broad groups—the metasediments, the basic plutonic rocks and the granites—but all are so distinctive and localized that it is logical to consider the complex as a unit (Fig. 3). Indeed the close spatial relationship of five granite bodies, all evidently related and apparently distinct from granitic bodies described elsewhere, serves to emphasize the unity of this small area of the crystalline basement.

The metasediments. These have been described by Hartley who was able to divide these ricks into pelitic and quartzitic groups, suggesting that a definite sedimentary sequence is present. Structurally these beds are disposed in an antiform plunging to the east—the Ballinderry Anticline. It is possible that this structure may be correlated with the third-fold phase of the Highlands, as at present understood, and if this structural correlation were

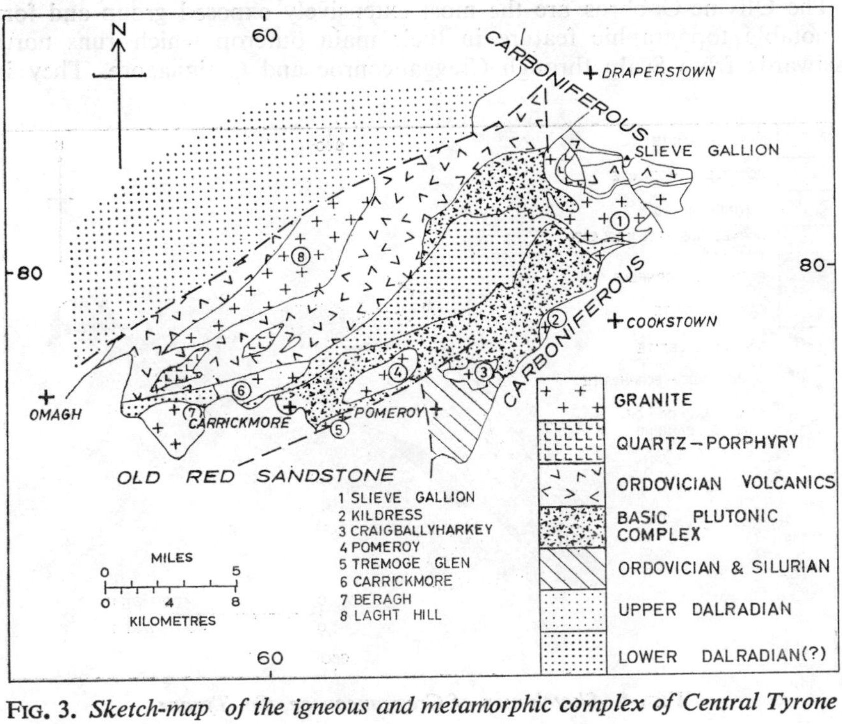

FIG. 3. *Sketch-map of the igneous and metamorphic complex of Central Tyrone*

upheld it would suggest that the metasediments are Dalradian or, at the oldest, Moinian. The folding and metamorphism of the metasedimentary inlier is complex. No detailed work has been done on it but at least three phases of folding can be recognized. The grade of metamorphism is high and the whole area lies within the sillimanite zone. Over most of the inlier the metasediments are principally mica-schists and psammites but at Corvanaghan quarry [H 718 812] a more migmatitic aspect is displayed.

The Basic Plutonic Complex. The basic intrusive rocks extend from Carrickmore to Lough Fea and can be subdivided into three groups:

i. The Early Dolerites and Gabbros
ii. The Olivine-Gabbros
iii. The Ophitic Dolerites of Carrickmore

The Early Dolerites are a very variable mixture of doleritic rocks intruded and injected by rocks of gabbroic grain size. They occupy an area immediately north of Carrickmore and north-eastwards, between the ophitic dolerite of Carrickmore and the olivine-gabbros. A narrow belt of the Early Dolerites on the south-east slopes of Cregganconroe [H 669 754] and Craignagore [H 677 762] is bounded on the south-east by the Pomeroy Granite, which has altered the dolerites to hornblende-hornfels and on the north-west by the Olivine-Gabbros, which have also altered the dolerites to pyroxene-hornfels (Fig. 4).

The Olivine-Gabbros are the most extensively exposed group and form a notable topographic feature in their main outcrop which runs north-eastwards from Scalp through Cregganconroe and Craignagore. They in-

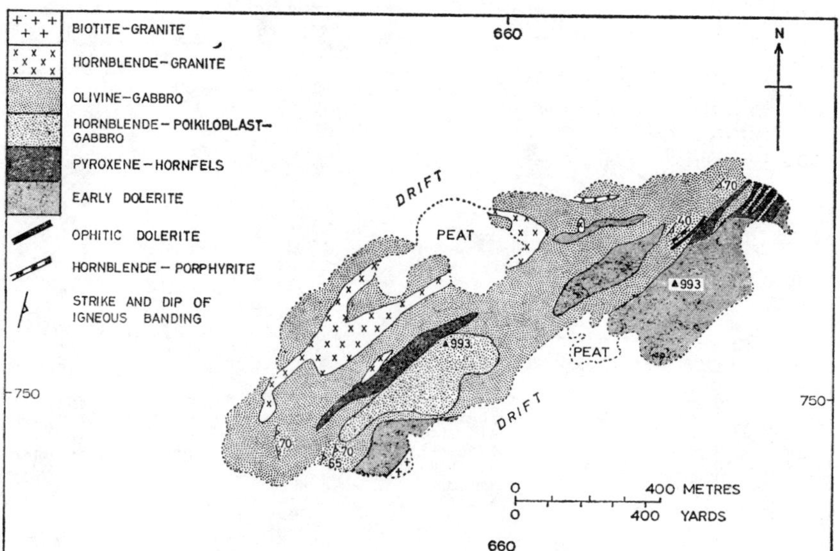

FIG. 4. *Sketch-map of Cregganconroe, Co. Tyrone*

trude the Early Dolerites and incorporate large rafts of dolerite at several localities. The gabbros are coarse pyroxene-plagioclase-olivine rocks, heavily uralitized, showing local banding as at Scalp Hill, and lenses of poikiloblastic hornblende-gabbro on the high points of Cregganconroe and Craignagore.

The late Ophitic Dolerites of Carrickmore occur only around the village, and though their intrusive contacts with the other groups are not exposed they are identical to dykes of ophitic dolerite which are seen to cut the Olivine-Gabbros. The rock is uniform and homogeneous and the ophitic structure can be seen in hand-specimens.

Structure within the Basic Plutonic Complex. The main unit boundaries follow a broadly Caledonoid trend but the igneous banding at Scalp Hill and elsewhere is oriented approximately north. Pockets of finely striped hornblende-schist are present on Scalp Hill and Craignagore and these are similarly oriented. The structure within these areas is complex and up to five phases of folding have been recognized. There can be no doubt that these pockets are actually portions of gabbro which became schistose. Foliated and schistose gabbros are also observed at dyke margins and it would appear that in general the whole basic plutonic complex was subjected to regional metamorphic stresses but was too massive to deform except at localities where a planar discontinuity existed. It is conjectural that the same regional metamorphism was responsible for the extensive uralitization of the gabbroic rocks and the metamorphism of the metasediments.

The Granites. The biotite-granites of Slieve Gallion, Kildress, Pomeroy, Carrickmore and Beragh are all so similar that it is not possible to distinguish them in hand specimens. The Pomeroy granite cuts the gabbroic rocks and hornfelses them slightly while the Carrickmore and Beragh granites, emplaced in metasediments, apparently barely alter the envelope material. In contrast the biotite-granite of Slieve Gallion is heavily contaminated with basic material from the envelope and in this respect is similar to granite dykes in Scalp Hill, Cregganconroe and Craignagore which are also heavily hybridized. Substantial areas on the western slopes of Scalp Hill have been permeated by granitic solutions.

Exposure is poor but such evidence as is available suggests that these biotite-granites were emplaced prior to the deposition of the Ordovician volcanic rocks which lie with apparent unconformity upon them.

The poorly exposed muscovite-microgranite of Tremoge Glen is cut by basic dykes of rock similar to the Carrickmore ophitic dolerite and may be earlier than the biotite-granites.

The granites of Craigballyharkey and Craigbardahessiagh, hornblende-tonalite and biotite-granodiorite respectively, both include blocks of the Ordovician volcanic series (Fig. 5) and are thus at least post-Arenig in age. Pebbles of these granites are said to occur in the Ashgillian sediments of Pomeroy and if this is true it would appear that the granites are of Ordovician age.

The biotite-hornblende-tonalite of Laght Hill is a medium-grained, highly sheared granitic rock. Because of extensive alteration it was not distin-

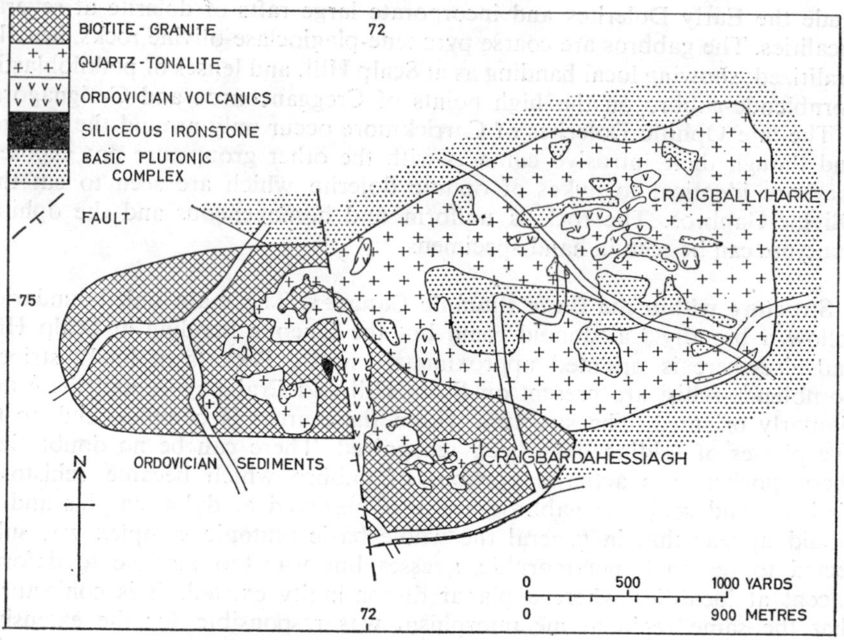

FIG. 5. *Sketch-map of Craigballyharkey and Craigbardahessiagh, Co. Tyrone*

guished from the enclosing Ordovician volcanic rocks by the original Survey. Hartley described it as sheared hybrid and recognized parts of it as grading into unsheared intrusive rocks, as for example of Cashel Hill. The age relationships of this tonalite are not clear but it is probably of Ordovician age.

Dalradian Assemblage

An area of about 2500 sq km in western Londonderry, Tyrone and north Fermanagh is underlain by schistose rocks which extend westwards to form most of Co. Donegal. In addition a small inlier of Dalradian rocks occurs in north-east Antrim.

Table 2 has been constructed using the format of that given by Johnstone (1966, based on Anderson 1948). In Northern Ireland most of the exposures are of Upper Dalradian. The Calcareous Sub-Group is relatively easily identified but the difficulty found higher in the succession in Scotland, within the Pelitic and Upper Psammitic Sub-Groups, will probably be repeated as detailed research continues in Northern Ireland. In Scotland these groups have been taken in the past to represent separate stratigraphical units but it is becoming increasingly difficult to maintain the division. Certainly in most of Ulster the probable equivalent rocks are not clearly grouped into pelitic and psammitic groups but are interbanded and it is possible that the boundaries of some are diachronous.

The order and relative ages of the Dalradian formations have long been the subject of argument, and through the development of 'way-up' criteria in the 1930's enabled successions to be established in several areas of the Scottish Highlands and Ulster it has been held for several decades that the successions of Islay and Perthshire (the Iltay succession) and Ballachulish (Ballappel succession) might be correlated laterally. Recent work in Islay, however, has shown that the Ballachulish succession is actually a downward continuation of the Iltay, and this relationship has been accepted in Donegal. The currently agreed successions in Donegal and Tyrone are given in Table 2.

The Moinian rocks of the Lough Derg area are cut off from the Dalradian rocks to the north by a structural break—the Lough Derg slide—and the district west of Castlederg is underlain by rocks high in the Dalradian succession, disposed in the form of an upfolded recumbent isocline—the Ballybofey Anticline. The core of this fold is made up of Lough Eske Psammites which are overlain by the Killeter Quartzite, the Aghyaran and Mullyfa Formations, the Shanaghy Green Beds, and the Croaghgarrow Formation—the Oughtadreen Pelitic Group of Anderson.

The axial trace of the Ballybofey anticline runs north-north-west, passing between Castlederg and Killeter, and on the south-western limb the beds are inverted so that rocks high in the Dalradian occur at the lowest structural level, next to and facing downwards on the Moinian.

There is considerable facies variation in the Upper Dalradian across the fold and the Mullyfa and Aghyaran formations are equated on the north-eastern limb by members of the calcareous Convoy Group.

The Dalradian rocks of the Sperrin Mountains are cut off to the south by the western extension of the Highland Boundary Fault, and no top or base to the succession is known. As described by Hartley they range from quartzites to graphitic phyllites and include a calcareous group which has been equated with the Loch Tay Limestone of the Scottish Dalradian succession (Table 2). Throughout the area the beds dip to the north-west and the groups of rock have linear north-east—south-west outcrop.

The succession in the south is probably inverted, and the oldest beds are apparently the Newtownstewart Quartzitic Formation of Lower Dalradian age (Middle Dalradian of Rast and Knill) in which colour-banded quartzites are interbedded with quartzose schists with thin beds of micaceous and black schists. The quartzites are most abundant in the lower part of the formation, which underlies a wide area on the north-west flank of the Sperrin Mountains and is well exposed on the high ground south-west of Newtownstewart and south of Loughan Hill [C 461 015].

The base of the Upper Dalradian is taken at the top of the succeeding Dungiven Limestone Group which consists of bands and lenticles of limestone with interbedded dark biotite-schists. The limestones vary in texture from pale saccharoidal to dark crystalline and in thickness from a few centimetres to many metres. Inadequate exposures make it impossible to say whether the variation is due to rapid lateral change or to the splitting up of individual beds. Where in contact with epidiorite intrusions the limestone is black and coarsely crystalline as at the old Priory Church at Dungiven [C 693 083]. The limestones can be traced from Pollangorm Hill, east of Dungiven, to near Newtownstewart and are probably equivalent to the limestones which outcrop around Raphoe in Co. Donegal.

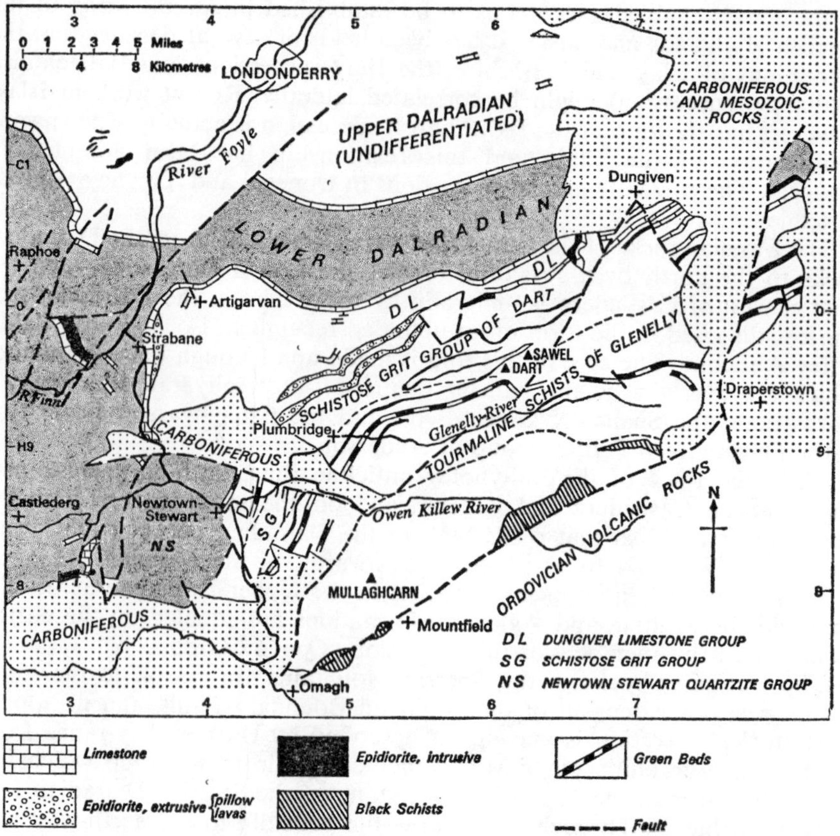

Fig. 6. *Sketch-map of the Dalradian rocks of the Sperrin Mountains* (After J. J. Hartley).

In the area north of Plumbridge the limestones are succeeded by dark green massive schists which outcrop in two bands nine or ten kilometres long and some hundred metres thick. Unlike the intrusive epidiorites these rocks, which vary from hornblende-schists to amphibolites, appear to have been of extrusive volcanic origin as they show well-marked pillow structure. This is best seen at the village of Craig [H 523 979] about eight km north-east of Plumbridge. Similar pillow lavas are seen at Strabane and Artigarvan to the west though it is not possible to say with certainty that they all lie on the same horizon.

Overlying the pillow lavas the Schistose Grit Group of Dart forms much of the higher ground of the Sperrins and includes pebbly grits and granulites as well as schistose grits. These beds are especially well seen near the summits of Dart and Sawel Mountains.

On the south side of the main Sperrin ridge the Tourmaline Schists of Glenelly overlie the schistose grits. The Tourmaline Schists are much more pelitic and contain abundant biotite and muscovite as well as the characteristic tourmaline which is best seen in the more micaceous bands. In the middle of the Tourmaline Schist Group a continuous band of albite-biotite-epidote-schist, equivalent to the Green Beds of the Dalradian rocks of Scotland, can be traced across country from near Newtownstewart to north of Draperstown. The thickness of this horizon in Tyrone is less than 30 m and the outcrop is rarely more than about 50 m across. In some localities it is associated with lenticles of impure limestones as at Oughtboy Burn, north of Cranagh [H 592 953] and at Garvagh Bridge [H 620 937]. It is generally believed that Green Beds are the result of the metamorphism of sedimentary rocks formed by the denudation of basic igneous material or deposited as pyroclastic tuffs.

The Tourmaline Schists are succeeded by the Mullaghcarn group of quartz-mica-schists, very variable in composition, which underlie a belt of country running north-east from Omagh. Albite-schists are very common in this group.

The highest beds seen in the Sperrin area are the Black Schists of Broughderg which crop out along the southern boundary fault. They are pelitic schists of a dull black or grey colour composed of quartz, graphitic dust and, in some cases, finely divided iron ore. They are well exposed in the Glenscollip and Glencurry Burns between Omagh and Mountfield and farther to the north-east in the Broughderg Burn.

The Sperrin succession, largely of Upper Dalradian age, can be closely correlated with the Inishowen peninsula rocks described by McCallian. The Culdaff and Inch Limestones are probably the equivalent of the Dungiven Limestone Group and by the use of sedimentary structures the sequence on Inishowen has been established to be generally right way up. South of Inishowen, in the vicinity of Londonderry City, the rocks are mainly phyllites, quartz-mica-schists and fine-grained schistose grits with only subordinate pebbly grits. These may be a continuation of the Inishowen Head Grits and Phyllites, preserved near the middle of the Irish equivalent of the Loch Awe Syncline.

Structure

The structure of the Sperrin area is assumed to be an overturned anticline with a generally inverted sequence dipping to the north-west. This may be a further continuation, from north-east Antrim, of the Cowal Anticline of Kintyre, part of the Southern Grampians nappe-complex. To the north of this major anticline there may be a generally synclinal structure, equivalent to the Loch Awe Syncline, centred on the area between Donemana and Inishowen (Fig. 7).

As well as the major overturned folds of counties Londonderry and Tyrone, probably the first major fold phase, two later stages have been proved. In the Londonderry area the first phase developed a strong E.N.E. axial-plane slaty cleavage. The second phase produced a N.W. strain-slip cleavage which, as in Scotland, frequently has schistosity corrugations with a smooth sigmoidal profile. Evidence for the third phase of folding and the resulting schistosity has been found on Inishowen. These major fold phases were followed by at least one, and probably more than one, episode of kink folding. Good examples of this can be commonly found in the Londonderry district, as at Creevagh Hill Quarry [C 402 158].

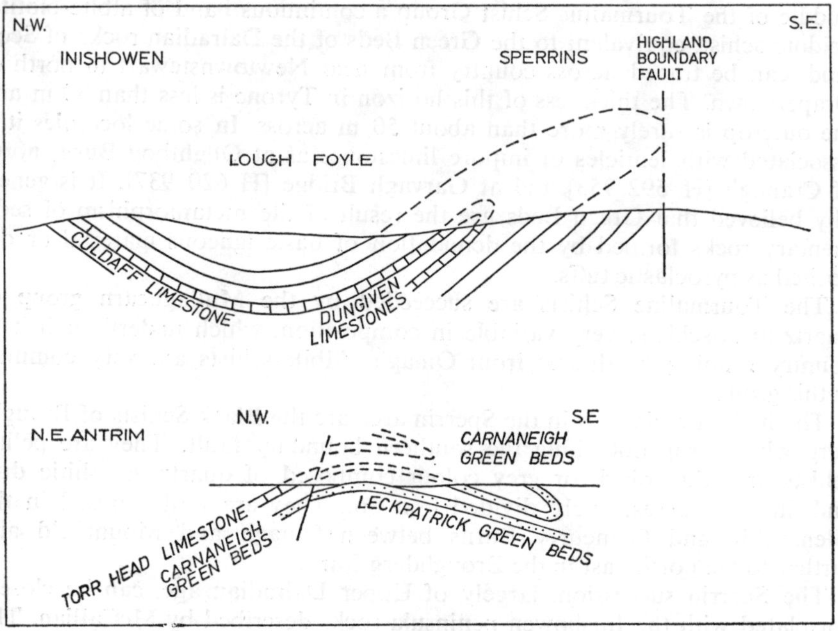

FIG. 7. *Conjectural structures of the Sperrin Mountains and Inishowen, and of north-east Antrim*

Antrim Inlier

In north-east Antrim, 50 km north-east of the Sperrin outcrop, an area of Dalradian rocks appears from beneath Cretaceous and Tertiary cover between Ballycastle and Cushendall (Fig. 8). The succession is predominantly of quartzose schists and schistose grits with one well-developed limestone group and Green Beds at two horizons and has been equated with the Upper Dalradian succession of Kintyre, only 21 km away across the North Channel.

The structure in Antrim is a continuation of that in Kintyre and Cowal, the exposed rocks forming the lower limb of the Cowal anticline. The succession seen is thus assumed to be inverted though there is little direct evidence of this, some inverted graded bedding at Runaby Head [D 256 370] being balanced by upward-facing grading at Port-aleen Bay [D 233 398].

The oldest beds seen, though structurally the highest, are the Murlough Bay Quartz-schists which form the cliffs and foreshore from Murlough Bay to Torr. The lower part of this group is markedly phyllitic and some beds of silvery grey phyllite are seen in inland exposures on the flanks of Knocklayd [C 108 388].

The Torr Head Limestone (Loch Tay Limestone) is well seen at the type locality, intimately associated with massive epidiorite. On the east of the headland it is a coarsely crystalline black marble, but it is usually grey and rather siliceous.

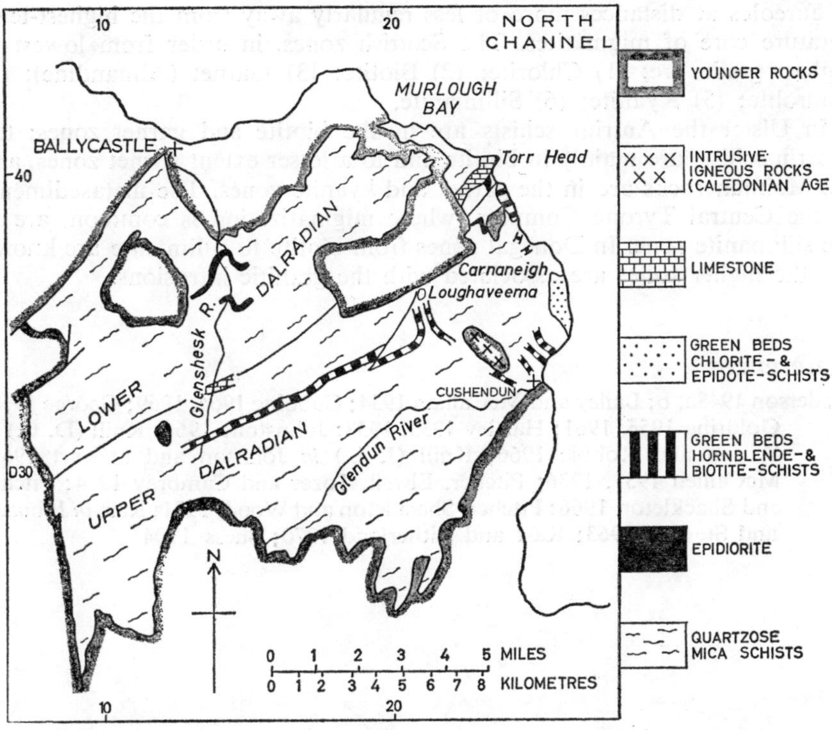

FIG. 8. *Sketch-map of the Dalradian rocks of north-east Antrim*

The overlying Altmore Schists are commonly crowded with phenocrysts of albite, as at Altmore Bridge [D 230 391], and are overlain in turn by the lower of the two Green Bed horizons, the pale green chlorite-epidote-schists, by the Loughan Bay Group of quartz-schists and grits with some Leckpatrick Green Beds [D 235 397], darker green hornblende-biotite-schists, by the Loughan Bay Group of quartz-schists and grits with some thin green bed and epidiorite horizons.

The youngest beds present are the Glendun quartz-schists and grits, equivalent to the Ben Ledi Grits, which are widely exposed in the southern part of the outcrop.

Metamorphism

The original sediments of the Moinian and Dalradian assemblages were subjected to dynamo-thermal metamorphism during the Caledonian orogeny, and considerable recrystallization of their constituents took place. Platy metamorphic minerals, especially micas, were developed with preferred orientations parallel to bedding or cleavage in the rocks, and acicular minerals grew with elongation commonly parallel to fold axes.

Variation in temperature, pressure, and the original composition of the rock caused the development of different assemblages of secondary minerals, and the work of Barrow and Tilley in the Grampian Highlands has led to the concept of metamorphic zones with index minerals arranged in aureoles at distances more or less regularly away from the highest-temperature core of migmatites. The Scottish zones, in order from lowest to highest grade are: (1) Chlorite; (2) Biotite; (3) Garnet (Almandine); (4) Staurolite; (5) Kyanite; (6) Sillimanite.

In Ulster the Antrim schists are in the biotite and garnet zones; the Sperrin schists are mainly in biotite and to a lesser extent garnet zones, and the Moinian rocks are in the garnet and kyanite zones. The metasediments of the Central Tyrone Complex, where migmatization is common, are in the sillimanite zone. In Donegal zones from biotite to sillimanite are known but the higher zones are associated with the granitic intrusions.

References

Anderson 1948a, b; Bailey and McCallian 1934; Cobbing 1964, 1969; George 1960; Goldring 1956, 1961; Hartley 1933, 1938; Johnstone 1966; Knill (D. C.) *in* Wilson and Robbie 1966; Knill (J. L.) *in* Johnson and Stewart 1963; McCallien 1931, 1936; Pitcher, Elwell, Tozer and Cambray 1964; Pitcher and Shackleton 1966; Pitcher, Shackleton and Wood, 1971; Rast *in* Johnson and Stewart 1963; Rast and Litherland 1970; Suess 1904

3. Lower Palaeozoic

The lowest system of the Palaeozoic, the Cambrian, is not proved in Ulster although, as in Scotland, part of the Upper Dalradian may be of Cambrian age. Ordovician and Silurian rocks, both volcanic and sedimentary, occur in counties Tyrone and Fermanagh and form an extensive belt of country extending from the Co. Down coast to Co. Longford.

Ordovician Volcanic Rocks of Tyrone

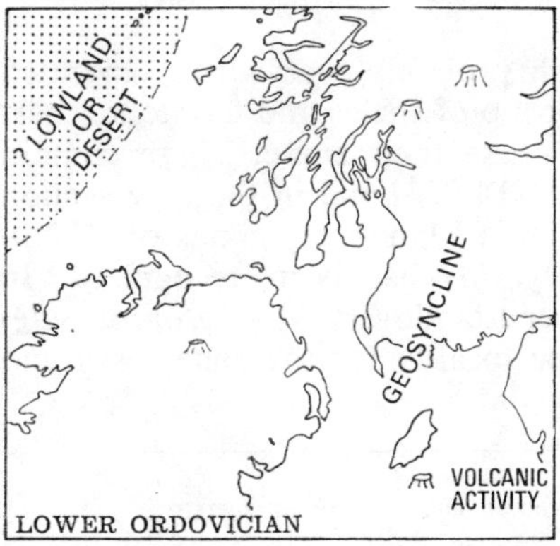

The volcanic series of mid-Tyrone was extruded on to an eroded surface of pre-Cambrian metasediments, rocks of the pre-Cambrian Basic Plutonic Complex and early biotite-granites. The eruptive rocks are mainly tuffs of andesitic composition, with subordinate spilitic pillow lavas and rare black shales and cherts, and were formed, at least in part, under water. Interbedded near the base of the succession is a red siliceous ironstone about 3 m thick which can be found at several localities across the outcrop at the same structural level.

By analogy with the volcanic rocks of the Highland Border series in Scotland—which like those of Tyrone are found along the line of the Highland Border Fault—and those of western Connemara, the Tyrone series might be assumed to be of Arenig age, but a graptolite fragment from a shale among the tuffs near the summit of Slieve Gallion has been identified as *Dicranograptus sp.*, probably of Caradoc age. It is possible that igneous activity persisted in this area through most of the Ordovician.

A number of quartz-porphyry intrusions, probably sills, are intruded into the volcanic series and are probably of the same general age.

Upper Ordovician and Silurian

Upper Ordovician (Ashgill and Caradoc) and Lower Silurian (Llandovery) rocks occur in two main areas; the Pomeroy inlier and the main Longford–Down outcrop.

The Pomeroy area is of interest as its abundant shelly fauna and varied rock-types indicate shelf conditions. The basal Bardahessiagh Beds are coarse grits, sandstones, and conglomerates, outcropping along the southern slopes of Craigbardahessiagh. Although at one time worked locally for flagstones these beds are now poorly exposed, but fossiliferous blocks can be found in walls. It was from these beds that excellent trilobites were collected when the quarries were working. The trilobite-brachiopod fauna is of uppermost Caradoc or low Ashgill age.

The overlying Killey Bridge Beds are a series of calcareous mudstones with a trilobite-brachiopod fauna of Ashgill age, well exposed in the stream north of Warren Wood [H 712 713]. The uppermost Ordovician rocks known, the Tirnaskea Beds, are calcareous mudstones, shales and sandstones and contain, in addition to a shelly fauna, the graptolites *Dicellograptus anceps, D. complanatus* and *Orthograptus truncatus,* indicating an Ashgill age. They are exposed in a stream near Slate Quarry Bridge [H 727 726].

Succeeding the Ordovician beds, though probably everywhere separated from them by faults, is the Little River Group of shales and flagstones, best exposed in the stream from which they take their name. These contain abundant graptolites at some localities [H 714 733] and in the type section a complete sequence of the lower and middle Llandovery zones of *Akidograptus acuminatus* to *Monograptus gregarius* has been recognized. In addition at Limehill [H 694 739] the upper Llandovery *Monograptus sedgwickii* Zone is present, this being the type locality for the zone fossil and for *Monograptus tenuis* (Portlock).

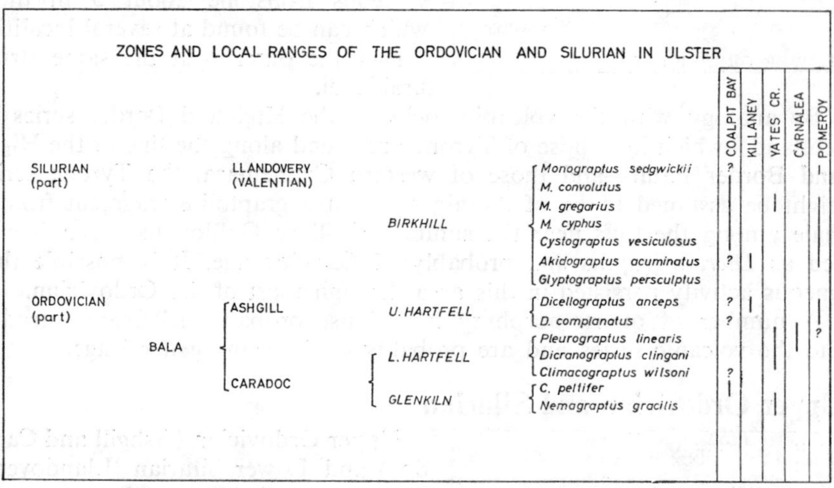

FIG. 9. *Zones and occurrences of the Ordovician and Silurian rocks in Ulster*

In the main Longford–Down outcrop both the Ordovician and Silurian rocks are largely of greywacke facies, formed in deep water mainly by submarine slumps or turbidity currents which carried large quantities of sandy sediments down the sides of the subsiding geosyncline. These greywackes (turbidites) are of sandstone grade and vary from a few centimetres to a few metres thick with a large proportion of rock fragments and fine-grained matrix. They are interbedded with thinner beds of siltstone or mudstone. There are a few occurrences in Co. Down of conglomerate with boulders of greywacke and metamorphic rocks, mainly quartzites. One well-exposed

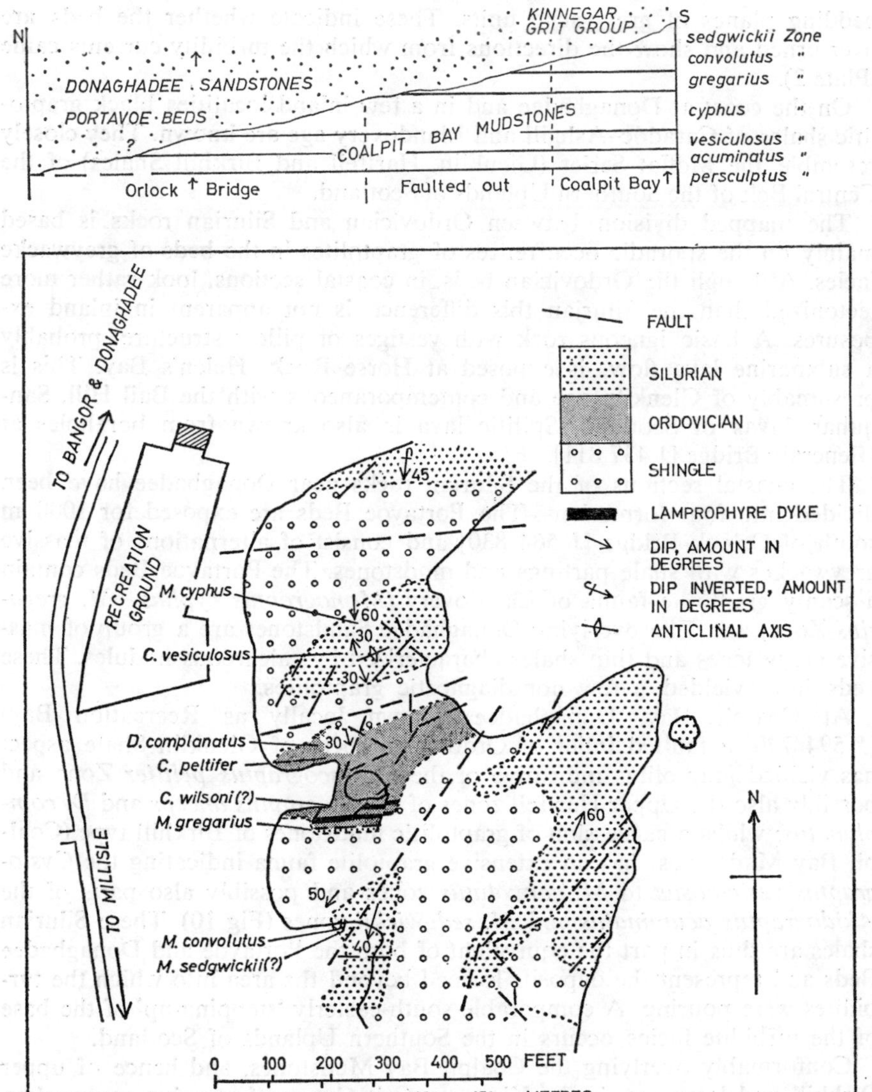

FIG. 10. *Sketch-map of Coalpit Bay, Donaghadee, with a section showing the regional lateral relationships of the Silurian rocks* (*see p. 22*)

outcrop at Ballyalicock [J 513 744] appears to be a slide-conglomerate of similar derivation to the turbidites.

The greywacke sequence is largely unfossiliferous, but a sparse shelly fauna, including corals and brachiopods, has been found. Graptolites are rare but a fauna of Glenkiln age is recorded from coastal exposures at Grey Point, Helen's Bay, and Lower Hartfell material from Carnalea. The turbidites are notable for the well-preserved 'sole-structures' seen on the basal bedding planes of greywacke units. These indicate whether the beds are overturned and show the directions from which the turbidity currents came (Plate 2).

On the coast at Donaghadee and in a few inland localities black graptolitic shales of Caradoc–Ashgill and Llandovery age are known. They closely resemble the Moffat Series (Glenkiln, Hartfell and Birkhill Shales) of the Central Belt of the Southern Uplands of Scotland.

The mapped division between Ordovician and Silurian rocks is based mainly on the sporadic occurrences of graptolites in the beds of greywacke facies. Although the Ordovician beds, in coastal sections, look rather more tectonized than the Silurian this difference is not apparent in inland exposures. A basic igneous rock with vestiges of pillow structure, probably a submarine lava flow, is exposed at Horse Rock, Helen's Bay. This is presumably of Glenkiln age and contemporaneous with the Bail Hill, Sanquhar, lavas of Scotland. Spilitic lava is also known from boreholes at Glencraig Bridge [J 437 811].

The coastal sections of the Silurian rocks near Donaghadee have been divided into four formations. The Portavoe Beds are exposed for 1000 m south of Orlock Bridge [J 564 830] and consist of alternations of massive greywackes with shale partings and mudstones. The Portavoe Beds contain a scanty graptolite fauna of Llandovery *(Monograptus cyphus—M. gregarius* Zone) age. The overlying Donaghadee Sandstones are a group of massive greywackes and thin shales characterized by calcareous nodules. These beds have yielded a few non-diagnostic graptolites.

At Coalpit Bay, Donaghadee (known locally as Recreation Bay) [J 594 790] a faulted inlier of Ordovician rocks of Glenkiln Shale aspect has yielded graptolites indicative of the *Climacograptus peltifer* Zone and possibly also the Upper Hartfell zones of *Dicellograptus anceps* and *D. complanatus;* whilst a succession of graptolitic mudstones of Birkhill type (Coalpit Bay Mudstones) has an extensive graptolite fauna indicating the *Cystograptus vesiculosus* to *M. convolutus* zones and possibly also parts of the *Akidograptus acuminatus* and *M. sedgwickii* zones (Fig 10). These Silurian shales are thus in part the equivalent of both the Portavoe and Donaghadee Beds and represent the deposits formed beyond the area into which the turbidites were pouring. A comparable south-easterly 'stepping-up' of the base of the turbidite facies occurs in the Southern Uplands of Scotland.

Conformably overlying the Coalpit Bay Mudstones, and hence of upper Birkhill and later age, is the Kinnegar Grit Group of massive greywackes. Excellent examples of sole structures—flute and groove-casts—are seen in these beds south of Galloway's Burn [J 595 784]. (Plate 2A).

EXPLANATION OF FIG. 11.

Ordovician. **1.** *Nemagraptus gracilis* (Hall); **2.** *Pleurograptus linearis* (Carruthers); **3.** *Orthograptus calcaratus* (Lapworth); **4.** *Dicellograptus anceps* (Nicholson).

Silurian. **5.** *Monograptus cyphus* Lapworth; **6.** *Dimorphograptus swanstoni* Lapworth; **7.** *Monograptus gregarius* Lapworth; **8.** *Akidograptus acuminatus* (Nicholson); **9.** *Monograptus convolutus* (Hisinger); **10.** *Monograptus jaculum* Lapworth; **11.** *Cephalograptus tubulariformis* (Nicholson); **12.** *Monograptus lobiferus* (McCoy); **13.** *Monograptus sedgwickii* (Portlock).

A. *Old Red Sandstone conglomerate, with cracked quartize boulders. Cushendun.*

PLATE 3

B. *Massive reef limestones of Viséan age. Knockmore Cliff.*

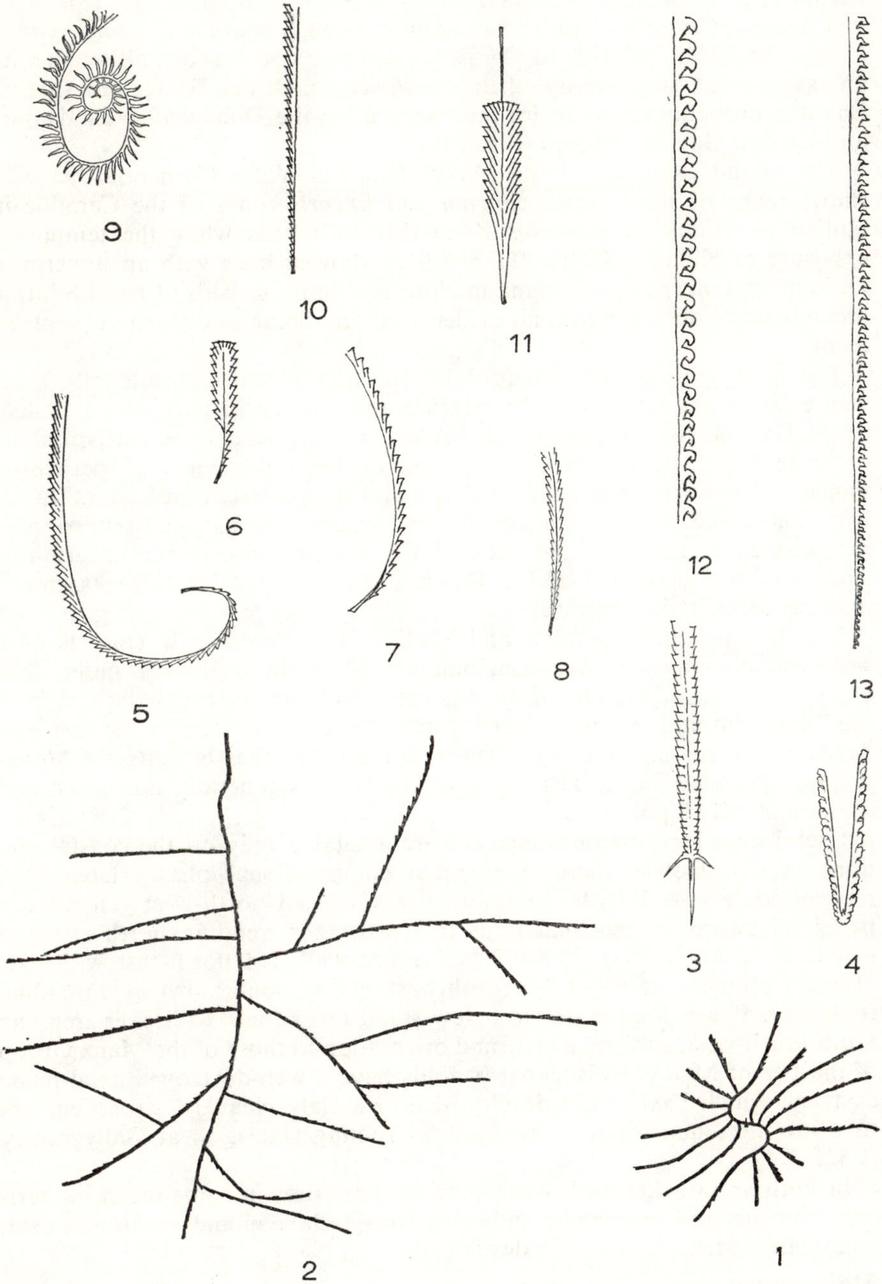

FIG. 11. *Characteristic Ordovician and Silurian graptolites*
(All drawings natural size, partly after Elles and Wood)

In the Southern Uplands of Scotland the base of the turbidite facies occurs at progressively younger horizons when traced south-east from Girvan *(cyphus* Zone) to Dobb's Linn and beyond *(maximus* Subzone, Upper Llandovery). The relation in Co. Down appears comparable, although with an exaggerated 'stepping-up' of the base between Orlock Bridge and Coalpit Bay due probably to the major fault separating the Donaghadee Sandstones and Coalpit Bay Mudstones (Fig. 10).

The inland exposure of the Moffat facies at Yates Corner [J 393 622] shows rocks of the *gracilis, clingani* and *linearis* zones of the Caradoc in faulted proximity to *gregarius* Zone (Birkhill) beds while the temporary exposure at Killaney House [H 366 613] showed beds with an uppermost Ordovician (*anceps* Zone) fauna in close proximity to beds of basal Silurian (*acuminatus* Zone) age with no evidence of any break in deposition between them.

The wide areas of the Longford–Down outcrop with a dominantly greywacke facies are assumed to be Silurian and may be in part the equivalent of the Hawick Rocks of Scotland. Isolated but individually extensive inliers of Ordovician rocks are either faulted or form the cores of periclines. Paucity of exposure and fossil evidence and the probability of lateral facies variation make any precise correlations impossible as yet. Bedded tuffs recorded from the Cootehill area of Co. Cavan, and, rather uncertainly, from the Portaferry area of Co. Down, are the only volcanic rocks known in the Silurian in this district.

The isolated Silurian inlier at Lisbellaw, Co. Fermanagh, consists of a series of mudstones with a conglomerate 50 m thick in their midst. The conglomerate has boulders of metamorphic and igneous rocks derived from the Dalradian and Arenig volcanic rocks which must underlie the Old Red Sandstone to the north, Graptolites from the mudstones indicate the *Monograptus gregarius* Zone. The strike of the beds is generally north-east and dips are usually high.

Useful data for structural analysis are available only on the coastal sections in Co. Down where three main compressional phases have been recognized. Phase 1 with maximum stress aligned north-west generated a periclinal swarm of isoclinal folds (F_1) with axes trending north-east and extensive strike-faulting. Phase 2 had a stress aligned north-east with axis of stress plunging at about 70° south-west, giving small north-west trending folds (F_2). Phase 3 gave rise to a few small cross-folds with axes trending north-north-west, similar in style and orientation to those of the Manx Group of the Isle of Man. The isoclinal F_1 folds have a well-developed axial-plane cleavage which has locally developed into a slaty cleavage strong enough to allow the mudstones to be used as roofing slates, as at Ballygrainey [J 525 794].

In both greywackes and mudstones the presence, in some areas, of sericite, chlorite and muscovite indicates low-grade regional metamorphism, and schistose texture is locally developed.

References

Cobbing, Manning and Griffith 1965; Fearnsides, Elles and Smith 1907; Harper and Hartley 1937, 1938; Hartley 1933; Manning, Robbie and Wilson 1970; Pollock and Wilson 1961; Sharpe, 1970; Swanston and Lapworth 1877.

4. The Caledonides

The Moinian and Dalradian rocks of the British Isles represent the western end of a thick belt of sedimentary rocks deposited on the Archaean basement in a long Caledonian geosyncline which stretched from Ireland to Scandinavia and beyond. Most of the sediments were Pre-Cambrian but deposition continued into the Lower Palaeozoic though to what extent is not clear. In Scotland it continued at least up to middle Cambrian. The earlier, Moinian, assemblage is characterized by uniformity of deposits through great thicknesses of sediments, probably accumulated in shallow water during slow subsidence at an early stage in the development, but the Dalradian assemblage is of diverse sediments probably accumulated in the more rapidly subsiding geosyncline proper.

In Connemara the Dalradian schists are overlain unconformably by Arenig beds of the *Didymograptus nitidus* Zone indicating that the metamorphism of the Dalradian took place at the end of the Cambrian, almost 495 million years ago. The first major episode of the Caledonian orogeny folded the sediments into a complex series of nappes and recumbent folds, similar to those now seen in the Alps, and metamorphosed the rocks into schists. The chain of fold-mountains thus formed were called by Suess the 'Caledonides'.

The folding was polyphase. The earliest stage gave major nappes with axes trending in general N.E.–S.E., and this was followed by secondary cross or transverse folds, generally on N.W.–S.E. or N.–S. axes. A third phase also occurred in many areas and fourth and later phases can be detected in minor structures. Movements of this type affected the Dalradian rocks of Ulster as well as those of Scotland, and in both areas the tectonic events were accompanied by major igneous activity.

Whether there was Cambrian sedimentation south of the Highland Border line in Ulster is uncertain, but in Lower Ordovician times part of the foreland to the south may have been land. From then on, however, the Southern Uplands geosyncline became the dominant feature and sedimentation into the subsiding trough continued till late in the Silurian. The ground-swell of the orogeny continued throughout this period with minor tectonic events reflected in varying conditions of sedimentation.

The end of Silurian times was marked by a second major orogenic event and the sediments in the trough were affected by compressional forces from the south-east. The rocks are highly folded, sometimes with the development of a strong slaty cleavage and commonly with accompanying mild (chlorite grade) metamorphism. The earlier idea of a vast 'anticlinorium' and 'synclinorium' of isoclinal steep-dipping folds has now yielded to the concept of large compound monoclines or asymmetrical anticlines made up of belts of tightly folded beds, in which the tangent to the fold crests is horizontal or inclined gently south, interspersed with zones of steeply dipping beds younging predominantly to the north-west. An imbricate series of reversed faults, hading north and trending north-east, brings up older beds to the north.

In addition to the folding and regional metamorphism the orogeny gave rise to a series of wrench faults with Caledonoid north-east direction, and to the two great fault-complexes which bound the Midland Valley in Scotland and run across Ulster—the Highland Boundary and Southern Uplands. The former probably enters Antrim about Cushendall and may be reflected in the Glenballyemon Fault. It is obscured by Jurassic and later sediments and lavas across Antrim and south Derry but reappears in Tyrone where it separates the Dalradian rocks of the Sperrins from Ordovician volcanic rocks and is a normal fault. West of Omagh it continues as the Castle Archdale Fault separating Dalradian from Old Red Sandstone and after disappearing beneath the Carboniferous west of Lough Erne, it continues westwards along the southern side of the Ox Mountains still below the Carboniferous.

The Southern Uplands Fault is not now seen at the surface, having been overstepped by Carboniferous and Triassic sediments, but its course is probably shown by the line of the Lower Palaeozoic massif south-westwards from Belfast Lough towards Longford, though its position must be some distance north-west of this line.

Many of the Caledonian faults were reactivated, sometimes in reversed directions, at later periods.

The main compressional phases of the orogeny were followed by the injection of a number of granite masses and minor acid intrusions.

The Newry Granite complex extends at outcrop for 42 km from Forkhill to Slieve Croob and has a maximum width of 10 km. The elongation of the exposure corresponds to the strike of the Silurian rocks and over much of the outcrop the contact is steep, with the country rocks dipping off at high angles. These steep and anomalous dips in the Silurian continue for 29 km to the north-east of Slieve Croob and 10 km south-west from Forkhill, suggesting an overall length of 80 km for the intrusion. The rocks of the complex vary considerably. Though usually a light grey, fine-grained biotite-granodiorite it locally has so much biotite and hornblende as to be dark in colour, as at Castlewellan. At its eastern end near Slieve Croob it is differentiated into biotite-peridotite, biotite-pyroxenite, shonkinitic monzonite, syenite, augite-biotite-diorite and hypersthene-monzonite (Fig. 12). Here, too, there appears to be a gradual transition from fused sedimentary rock into igneous material and this granite was regarded by the original Geological Survey and by later workers, as migmatitic in origin—formed by the melting of the original sediments. The main body of the granodiorite often shows a marked fluxion texture parallel to the contact and the strike of the Silurian, and has inclusions of Silurian material with bedding parallel to that of the country rocks. These factors have also lent weight to the migmatitic hypothesis though they may be due to magmatic flow and provide an example of structural control of intrusion. The Silurian mantle to the granite is highly altered to hornfels for considerable distances from the contact, and shows signs of thermal metamorphism in an aureole up to a kilometre wide around the outcrop. It seems probable that the mass was a high-temperature intrusion which has partially assimilated the envelope in some areas.

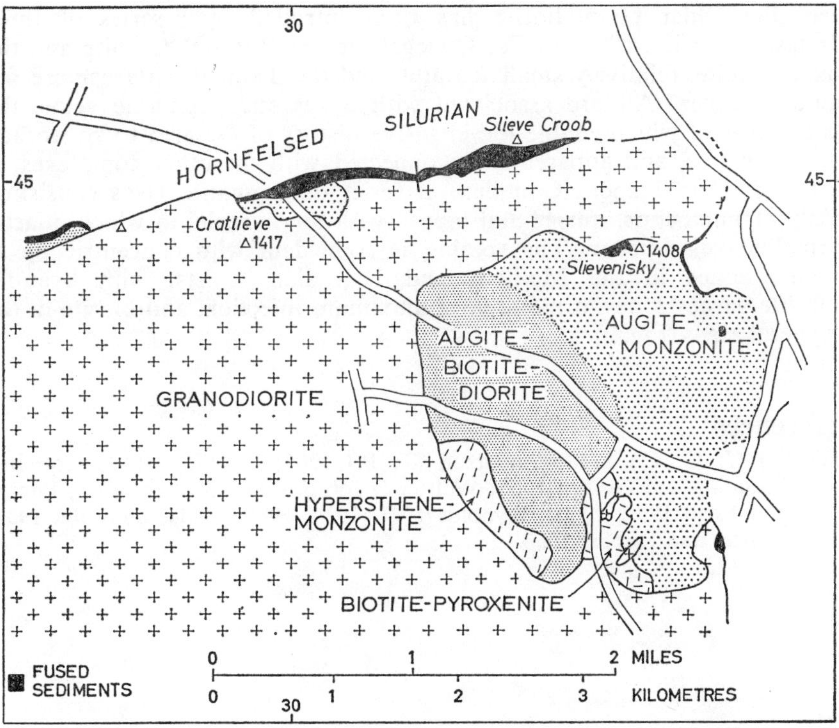

FIG. 12. *Sketch-map of the north-east end of the Newry granodiorite* (After D. L. Reynolds).

Associated with the Newry Granite is a north-easterly dyke swarm, best displayed on the coasts of the Ards peninsula. These intrusions are divided into two series—an older, intruded during a pause in the compressive earth movements and subsequently sheared and crushed, and a younger, intruded after the end of the folding. The older series are confined to the southern part of the Ards peninsula, while the younger lie north of a N.W.–S.W. line through Cloghey. The older group are all lamprophyres, mainly pyroxene-minettes, but the younger include feldspar-porphyry, hornblende-porphyry, minette, kersantite, vogesite and spessartite.

The granites associated with the igneous and metamorphic complex of central Tyrone are Caledonian in age and have already been described (pp 9–10). A small granite intrusion at Crossdoney near Cavan is probably of Caledonian age, as are the small granite boss and minor intrusions of pink quartz-oligoclase-porphyry near Cushendun. There are two small sills of lamprophyre in the Dalradian schists in the Torr Head area.

The Donegal granites are numerous and extensive and can only be mentioned briefly here. The main Donegal Granite is the last of four major plutonic phases in that area; an older Migmatitic Granodiorite which forms the north-western seaboard of the country around Bloody Foreland: the Ardara pluton which was emplaced in a schist 'envelope' and consists of a granodiorite core with a tonalite sheath; and the Rosses Ring-complex, where the Older Granodiorite has been intruded by a series of three granites. Other intrusives in Co. Donegal are the Rosguill Granite and the Fanad Granite, relatively small features, and the distinctive Barnsmore red granite complex. All are associated with dykes and pegmatite veins, the relative ages of which enable the intrusive history of the area to be worked out. Distinctive, and apparently unconnected with the other complexes, is the Lough Derg Igneous Complex—a series of pegmatite veins consisting mainly of microcline, quartz and biotite with an 'aureole' in which quartz-tourmaline veins occur. The veining is coincident with concentric metamorphic zones in the Moinian and Dalradian schists, with kyanite-staurolite-garnet zones in the area of maximum injection, and biotite in the peripheral areas (Fig. 2).

References

Anderson (J. G. C.) 1948; Anderson (T. B.) 1965; Cobbing 1964; Cobbing, Manning and Griffith 1965; Walton *in* Johnson and Stewart 1963; Pitcher and Read 1959, 1960; Reynolds 1931, 1934, 1936, 1943a, b; Skiba 1952; Wilson and Robbie 1966.

5. Old Red Sandstone

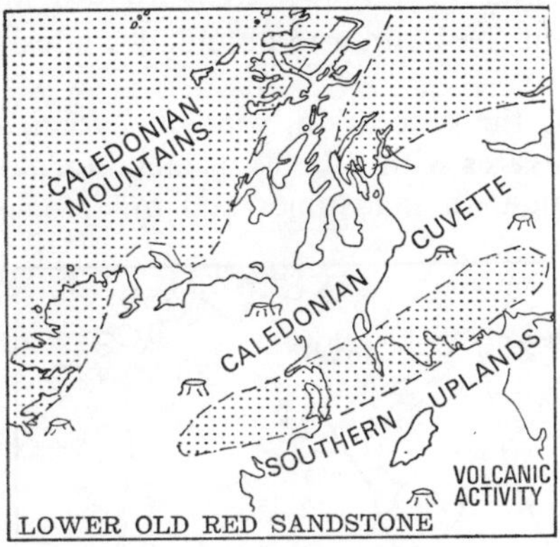

LOWER OLD RED SANDSTONE

The mountain-building effects of the Caledonian orogeny left the north of Ireland and Scotland dominated by the metamorphic-rock Dalradian uplands which towered over the intensely folded foreland of Ordovician and Silurian rocks at the beginning of the Old Red Sandstone period. Far to the south Devonian marine sediments were being deposited in the sea which bordered this foreland, but the wide continental area was dry land with a generally dry climate, interrupted by occasional periods of heavy rainfall. The high, newly formed mountain chain was susceptible to the attacks of the elements, and seasonal torrents carried vast quantities of debris down to the plains to the south and into the north-east to south-west valleys of the fold mountains. Much of this material was deposited as torrent fans of unsorted detritus which now form the massive conglomerates of Dalradian rock debris typical of much of the Lower Old Red Sandstone. Sometimes lakes or inland seas were formed, generally short lived, and in these the sediments were water-sorted so that beds of sandstones, mudstones and shales were deposited, often with the mud-cracks and ripple marks typical of shallow water deposits. Rare fish remains have been found in some of these sediments, and *Pteraspis* fragments from near Lisbellaw are the only known fossil evidence for the age of these beds in Ulster.

In some areas the last phase of the Caledonian igneous period gave outbreaks of volcanic activity with the eruption of tuffs and andesitic lavas.

Old Red Sandstone rocks crop out over 750 sq km of Counties Fermanagh and Tyrone, from Lough Erne to the region of Pomeroy. The Old Red Sandstone is cut off on most of its northern flank by major faults and its base is seen only near Pomeroy, where it overlies the Silurian inlier. It is in turn unconformably overlain by Carboniferous rocks to the south and east. The sediments are commonly conglomerates but there are extensive developments of sandstones and subordinate mudstones. Contemporaneous andesitic lava flows occur at several levels.

The outcrop is bisected by the Tempo–Sixmilecross Fault, a north-east trending dextral wrench with a postulated lateral movement of 20 km. In the eastern sector localized basal conglomerates overlying the Silurian near Pomeroy are succeeded by a predominantly sandstone group at least 600 m thick. The sandstones are reddish brown and green in colour and are locally flaggy. Typical beds can be seen at Gortnagarn Bridge [H 694 718] near Pomeroy. The succeeding andesite lavas may be 500 m thick and are well displayed on Barrack Hill [H 695 675]. They are purplish brown rocks and often show flow-banding and magmatic rolls.

Overlying the lavas is a considerable thickness of massive conglomerates which form the high ground north of Ballygawley. The conglomerates are largely composed of greywacke and andesite pebbles with an almost complete absence of quartz or metamorphic rocks. In the scarp at Knockmany [H 543 555] good sections in a ravine show the proportions of these components changing from greywacke alone at the bottom to abundant andesite at the top of the scarp.

West of the Tempo–Sixmilecross Fault the succession, probably younger than that to the east, consists of conglomerates with pebbles of metamorphic rocks interbedded in rhythmic succession with sandstones and mudstones.

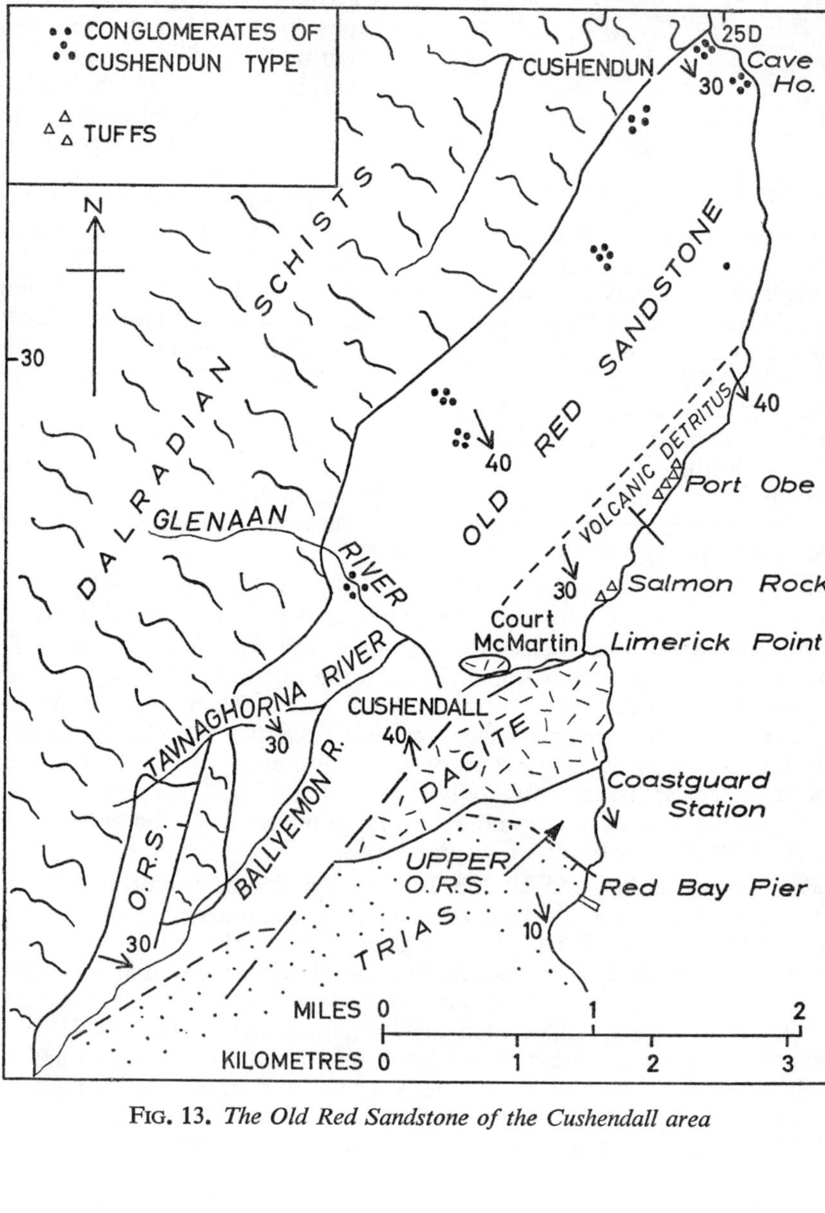

FIG. 13. *The Old Red Sandstone of the Cushendall area*

The proportions of these components vary throughout the succession so that while the lowest rhythms seen are predominantly sandstones, succeeding groups are dominated by conglomerates, sandstones and mudstones, in that order.

The sandstones and mudstones have normal sedimentary structures—current bedding, ripple marks, mud-cracks, load-casts, etc.—and the whole succession is dull purplish or reddish brown in colour, though lighter beds of sandstone occur. The rhythmic sequence is particularly well exposed in a quarry at Largy [H 298 474], 5 km west of Tempo, where false-bedded feldspathic sandstones have several conglomerate beds. Mudstones with mud-cracks are seen in a road cutting at Doogary [H 467 698] near Omagh.

Baryte occurs widely throughout the outcrop, both as a cement in the arenaceous beds and as veins. It is particularly noticeable near the Tempo–Sixmilecross Fault.

Old Red Sandstone rocks also outcrop in the area round Cushendall, where they rest unconformably on Dalradian schists (Fig. 13). The lowest beds are massive conglomerates containing a high proportion of quartzite, in rounded and cracked cobbles, which is not now known at outcrop anywhere among the Dalradian rocks of the district but which is similar to the Islay quartzites (Plate 3A). The conglomerates and some interbedded sandstones and siltstones are probably about 90 m thick and are overlain by some 900 m of red and purple sandstones with ripple-marks and mud-cracks. These pass upwards into increasingly conglomeratic beds containing an increasing proportion of andesite debris. In the highest exposures these conglomerates consist entirely of andesite boulders and are interbedded with coarse tuffs which contain andesite blocks up to 2·5 m in diameter, well exposed on the foreshore at Salmon Rock [D 243 283].

The source of this volcanic material may have been a vent in the village of Cushendall which is now represented at Court McMartin by a plug of quartz-andesite. An intrusive sheet of similar rock crops out south of the river and is overlain by a small area of conglomerate containing boulders of andesite, metamorphic rocks, and of an older conglomerate. This deposit is probably of Upper Old Red Sandstone age but precise dating is not possible.

Far to the north of the Highland Boundary Fault a small faulted outlier of Old Red Sandstone conglomerate occurs at Ballymastocker, Co. Donegal. While the Tyrone and Antrim occurrences are both part of the old foreland deposits, that at Ballymastocker may be the relic of an isolated valley deposit.

References

Andrew 1951; Bishop 1951; Harper and Hartley 1938; Nolan 1880; Wilson 1953.

6. Carboniferous

At the close of the Old Red Sandstone period the sea lay far to the south and during the slow advance of the Tournaisian transgression across the southern part of the British Isles erosion of the Old Red Sandstone, and probably the deposition of very similar sediments from the debris, proceeded in Northern Ireland and Scotland. When the marine transgression reached this area is uncertain but there is some evidence that it arrived in the late or even mid-Tournaisian period and covered a peneplain on which the rocks ranged from Moinian to Old Red Sandstone in age.

Unlike Scotland, where the line of the Highland Boundary Fault marked the edge of a highland massif beyond which the sea encroached only to a limited extent, the line of this fault in Ulster seems to have been of less importance than the regional down-warping of the area north and west of the old Down–Longford massif and the Ox Mountains anticline. Gentle crustal flexing seems to have gone on during much of the Carboniferous period. There is some indication that in the west the Highland Boundary Fault was actually reversed, with downthrow to the north-west, at this period, and the greatest thicknesses of Carboniferous rocks were deposited in this area.

Basal Clastic Group

The earliest deposits of the Carboniferous sea were of deltaic facies, laid down in a subsiding basin on the flanks of a land area to the north-west. The resultant boulder beds, grits and sandstones are notable for their impersistence and their rapid lateral changes of thickness. They are apparently diachronous, the area of deposition moving north with the slow advance of the marine transgression. In south Tyrone and Fermanagh they are of Tournaisian age; in Donegal and east Tyrone they are Lower Viséan (C_2S_1–S_2) and in Derry they are Upper Viséan (D).

As the transgression moved northwards across the metamorphic rocks and Old Red Sandstone the coastal deposits gave place in deeper water to silty shales and limestones, the latter often in the form of thin ribs in a shale succession. Occasional beds of sandstone indicate an intermittent return to deltaic conditions, presumably caused by continuing crustal instability.

The regional downwarp in which the great thicknesses of the deltaic facies accumulated was a belt running north-east from Ballina through Sligo towards Omagh and Newtownstewart. Post-Carboniferous folding and denudation has left a series of isolated or semi-connected synclines—Ballina,

Sligo, Donegal, Omagh, and Newtownstewart—in which these deposits are preserved.

In the Donegal syncline, sandstones and conglomerates, over 1100 m thick in the area north-east of Donegal Bay, pass southwards into the Ballyshannon Limestone (of probable C_2S_1 age) which rests on the basement rocks with only a thin conglomerate at the base. In the Omagh syncline the great thickness of sandy beds is divided into a basal series, the Omagh Sandstone Group, 600 m thick in the type area and very conglomeratic, which is succeeded by the 1100 m thick Claragh Sandstone Group. This group oversteps the Omagh Sandstones on to the metamorphic rocks to the west, where a basal conglomerate rests on the Dalradian rocks. The group is calcareous at the base and top and sandy or arkosic elsewhere. This great thickness of over 1500 m of sandy sediments, probably of Tournaisian age in part, is overlain by the Pettigo Limestone, marine limestones of C_2S_1 age. In the isolated Newtownstewart basin to the north only sandy beds are known and they may be as much as 2400 m thick.

In the Clogher Valley conglomerates and sandstones at least 300 m thick flank the Old Red Sandstone north of Fivemiletown, where they are overlain by Tournaisian limestones near Cole Bridge [H 443 523], and eastwards towards Ballygawley arenaceous and silty beds underlie a wide area and must be of at least this thickness. They are well exposed in stream sections at Lisdoart [H 618 555] and Lisbeg [H 639 565]. North-east of Ballygawley massive conglomerates pass upwards into a group of white sandstones 30 m thick and the total thickness of the deltaic and coastal beds here is probably 200 m. The overlying limestones are of S_2D_1 (Viséan) age and, unless the whole Lower Viséan succession is cut out by a non-sequence or fault, the clastics must be of Viséan age.

In the Cookstown area the basal Derryloran Grits, 140 to 200 m thick, are a series of red fine-grained conglomerates (which include limestone pebbles), pebbly grits, and quartzose sandstones, and include a 3 m sandy dolomite and dolomitic nodules in some of the sandstones. Good exposures of the dolomite and sandstones occur in the Ballinderry River [H 805 768]. These beds are probably of C_2S_1 age, though a cerioid *Lithostrotion* from Kildress suggests a slightly later age.

The overstep of the Carboniferous beds onto the Down–Longford massif probably did not occur till later than the transgression northwards—perhaps not till upper Viséan times. East of Armagh sandy and occasionally conglomeratic beds rest on the Lower Palaeozoic rocks, and reach a thickness of some 150 m [H 894 463]. To the south-west the arenaceous facies appears to thin out and south of Monaghan the limestones, sometimes sandy, are thought to rest directly on the greywackes. Sandstones and grits appear again round Cavan and thicken to the west and in the Drumod area there are extensive outcrops of conglomerates.

The Old Red Sandstone inlier of the Curlew Mountains is also overlain by conglomerates and sandstones of Carboniferous age, apparently some hundred metres thick, on its eastern and southern flanks.

In the same stratigraphical position, overlying the northern flank of the Lower Palaeozoic massif, there is a small outlier of Carboniferous rocks at Cultra, on the southern shore of Belfast Lough. Here about 180 m of

ZONES AND LOCAL DIVISIONS OF THE CARBONIFEROUS IN ULSTER							
Series	Coral - Brachiopod Zones	Goniatite Stages	Non-marine Lamellibranch Zones	Lithostratigraphical Divisions	W. Fermanagh	E. Fermanagh and Tyrone	North Derry
LOWER WESTPHALIAN (Ammanian)		*Gastrioceras* (*G*)	*Anthraconaia lenisulcata*	COAL MEASURES		Coal Measures	
NAMURIAN C				Missing or thin			
NAMURIAN B		*Reticuloceras* (R_2) (R_1)					
NAMURIAN A		*Homoceras* (*H*) *Eumorphoceras* (E_2) (E_1)		MILLSTONE GRIT	Millstone Grit Cuilcagh Shales	Millstone Grit Rossmore M'st.	Sandstone and thin dolomite
VISÉAN	*Dibunophyllum* (D_2) (D_1)	*Posidonia* (P_2) (P_1) *Beyrichoceras* (B_2)		UPPER CLASTICS UPPER LIME-STONE	Dartry L'st.	Slieve Beagh Clastics Gp. Upper L'st.	Basal sandstone and thin coals
	Seminula (S_2)	" (B_1)		CALP	Glencar L'st. Benbulben Shale Mullaghmore Sandstone Bundoran. Shale	Calp shales, sandstones and limestones	
	Upper Caninia (C_2S_1)	*Pericyclus* (*Pe*)		LOWER LIME-STONE	Ballyshannon Limestone	Lower Limestone	Cultra Co. Down
TOURNAISIAN	*Lower Caninia* (C_1) *Zaphrentis* (*Z*) *Cleistopora* (*K*)			BASAL CLASTICS		Basal sandstone shales, cement stones and limestones	Shales and cementstones. Basal sandstones

FIG. 14. *Zones and local divisions of the Carboniferous rocks in Ulster*

red sandstones with some shale bands are overlain by grey shales with cementstone ribs and some thin fossiliferous limestones near the top. The total thickness exposed is probably over 300 m. Recent work on ostracod faunas from the upper beds suggests that they are middle Tournaisian, as are the Cementstones of Ayrshire. It is likely that the Cultra basin was separated from the areas to the west by a pre-Carboniferous ridge running northwards and that deposition here in a marine environment with restricted circulation, perhaps lagoonal, was much earlier than anything elsewhere exposed in Ulster except the deltaic facies in the west.

North of Cookstown present outcrops show an irregular strip of Carboniferous rocks extending northwards to Dungiven and Londonderry, between the Dalradian to the west and the overlying Mesozoic rocks to the east. This limited exposure is the only evidence left of the further encroachment of the Viséan transgression across the basement to the north. These beds are mainly of deltaic facies, consisting of reddish and pale coloured sandstones and conglomerates with subordinate shales and thin coals. It seems possible that they once extended over the Dalradian rocks to the west, and probably all of Tyrone and Londonderry were included in the transgression.

In a borehole at Magilligan arenaceous rocks with two thin coals are shown by palynological dating to be very late Viséan. To the east nothing is known about the extent of Carboniferous beds below the Antrim plateau except that they are present at Ballycastle, (p. 40), are absent over the Dalradian ridge between Ballyvoy and Cushendun, and are absent at Langford Lodge. It seems a reasonable assumption that they once covered much of this area but that they were folded and largely eroded before Trias times. Relict basins may occur beneath the Mesozoic cover particularly, as Wright suggested fifty years ago, in the Coleraine/Ballymoney and Larne areas.

Viséan Limestones

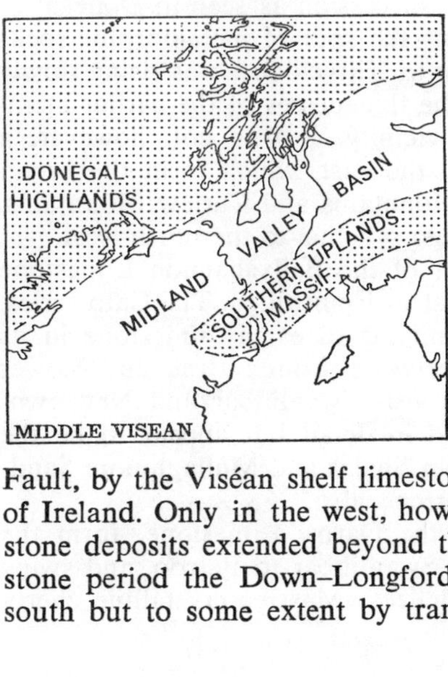

On the maps of the Geological Survey of Ireland this group of deposits was divided into Lower Limestone, Calp, and Upper Limestone, the Calp being itself divided on occasion into Calp Shale, Calp Sandstone and even Calp Limestone. Although the subdivision of the Calp is often of only local significance the group itself is valid as a largely argillaceous and arenaceous succession which is recognized over wide areas.

The basal deltaic facies was succeeded, south of the Highland Boundary Fault, by the Viséan shelf limestones which cover a wide area in the south of Ireland. Only in the west, however, is there any evidence that the limestone deposits extended beyond the Highland Boundary. During the limestone period the Down–Longford massif was overridden mainly from the south but to some extent by transgression from the north. It was largely covered with Carboniferous deposits, most of which have now disappeared leaving only the outliers of Carlingford, Stradone, Strangford, and Kingscourt, probably preserved by down-faulting. It is probable also that the Ox Mountains and Curlew Mountains anticlines in the west were submerged.

In the Sligo and Ballyshannon areas the lower part of the limestone succession consists of over 300 m of massive limestones of C_2S_1 age with few and thin shale partings. As noted earlier there is a replacement of these limestones by sandier beds to the north and beyond Lough Eske they are entirely represented by arenaceous sediments.

This Ballyshannon Limestone is succeeded, north of the Erne, by the Coolmore Shales and Kildoney Sandstone 150 and 90 m thick respectively. Further south, in the area between Sligo and Lough Melvin, the full Calp succession is represented by the Bundoran Shale, 140 m, the Mullaghmore Sandstone, 180 m, and the Benbulbin Shale, which is 90 m thick and contains an increasing proportion of limestone bands in its upper part. These are succeeded by the Glencar Limestone, 180 m thick, very shaly in its lower part and probably of D_1 age, which is the 'Calp Limestone' of the older description, and the Dartry Limestones—the Upper Limestones of the one-inch maps—which are over 210 m thick and consist of massive granular crinoidal limestones with abundant corals and practically no argillaceous or sandy debris. Chert is abundant in these beds and there is a notable development of bryozoan reefs in the form of lenticular masses of calcite mudstone several hundred metres thick and covering several square kilometres.

In the Omagh syncline the Pettigo Limestone, about 300 m thick and of C_2S_1 age, is the equivalent of the Ballyshannon Limestone to the west. It consists of massive highly fossiliferous limestones, locally bituminous and occasionally oolitic. Thin shale partings become more common in the upper part of the group. Here the limestone succession is abruptly but apparently conformably succeeded by the Clonelly Sandstone Group, over 450 m thick, without an intervening shale succession as seen in Donegal.

The Highland Border ridge which forms the spur of metamorphic rocks running from the Ox Mountains north-east to Kiltyclogher was rising slowly during early Viséan times, and reduced the thicknesses of the earlier members of the limestone succession in its vicinity. It was not, however, a barrier to transgression and the beds to the east, which form the broad Carboniferous syncline between the Ox Mountains and Curlew Mountains, are very closely similar to those in the Sligo syncline to the north-west.

The Lower Limestones—the equivalent of the Ballyshannon Limestone—are seen along the low ground north-west of Enniskillen. The Calp series, shales and sandstones with increasing amounts of earthy limestone in its upper division, underlies a wide area between Lough Erne and Lower Lough Macnean and south-eastwards towards Swanlinbar and Newtownbutler. These beds have not been studied in detail but appear to be the equivalents of the Bundoran and Benbulbin Shales and Mullaghmore Sandstone and possibly of the Glencar Limestone also.

The Upper Limestones, equivalent of the Dartry Limestone, form the scarps which flank the high ground from Swanlinbar to Belcoo, and swing in a great arc west of Derrygonnelly to Belleek. Massive crystalline cherty

PLATE 4

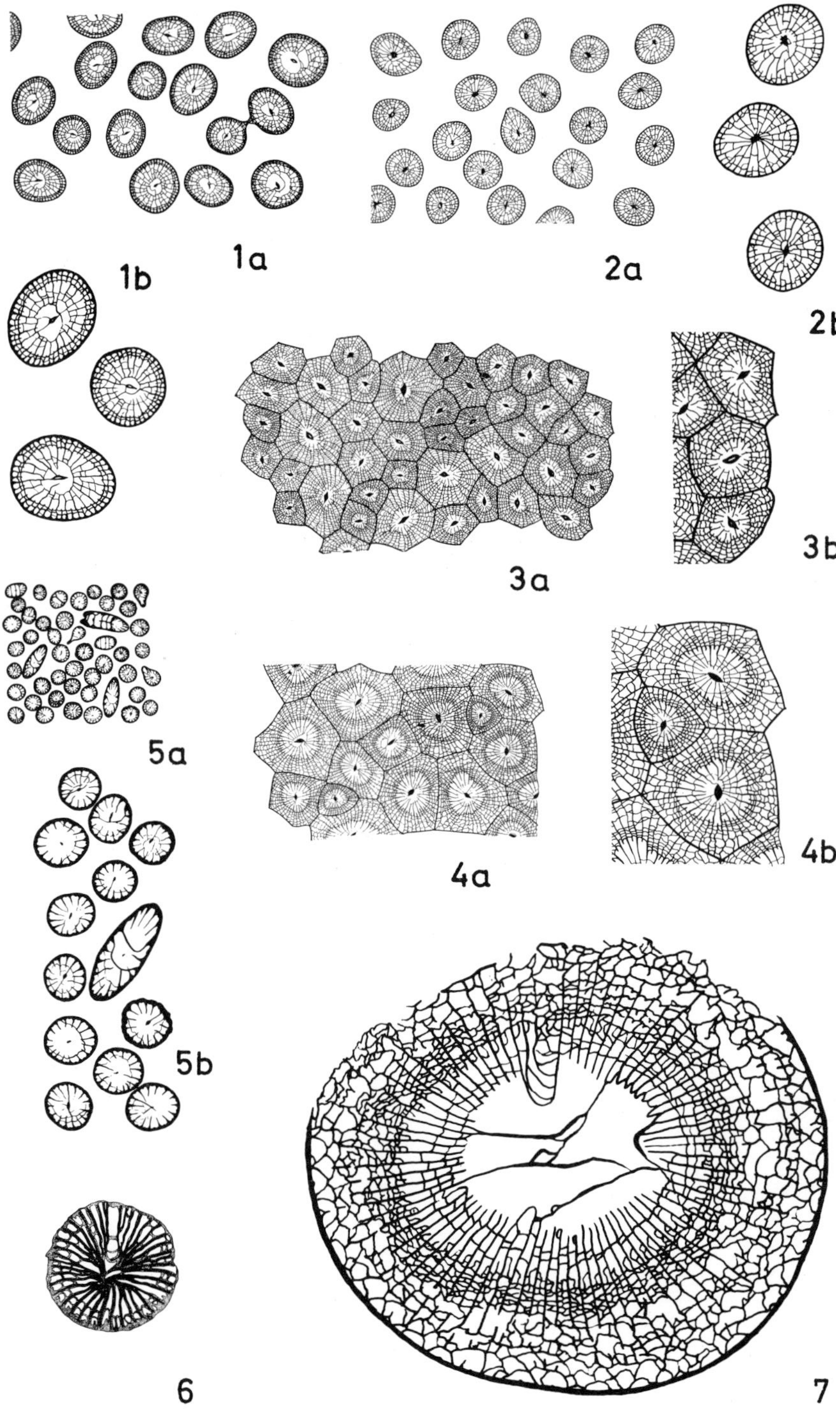

Some characteristic Lower Carboniferous corals

limestones, locally riddled with swallow holes and underground streams, are over 200 m thick south-east of Belcoo. They include large unbedded reefs, like those of the Sligo syncline.

In the Carrick-on-Shannon area, some 30 km to the south-west, the succession is completely different. The basal clastics are succeeded by 900 m of thin-bedded limestones and shales alternating with massive crinoidal limestones. Lenticular sheet reefs are interbedded with the Upper Limestones. This series ranges from S_1 to D_1 age. Compared to the Sligo-Lough Erne area the succession is thinner and devoid of arenaceous beds and thick shale groups and marks a transition from the north-west facies to the shelf-limestone facies of the central plain.

East of Lough Erne the Viséan succession is about 2000 m thick. The Lower Limestone is argillaceous towards the base but the upper part of the group (C_2S_1) is of clear stylolitic biomicrites, seen in quarries at Crieve Hill [H 405 483]. These are succeeded by a great thickness of 'Calp' (S–D_1)—earthy or argillaceous limestones and calcareous siltstones with at least one thick group of sandstones, well exposed east of Aughnacloy—and Upper Limestone (D_1) which is massive micrite with some shales, seen in the Glenoo area, south-south-east of Fivemiletown. There is little chert in these beds and reef-facies development is scanty, known only from the Plaister area [H 678 540] north-east of Aughnacloy.

The Calp lithology can be traced north-eastwards from Aughnacloy almost as far as Stewartstown. Near Dungannon the sequence includes two thick sandstones, formerly quarried for building stone near Carland and Newmills.

Towards Cookstown the argillaceous Calp passes into regularly bedded limestones with a few thick sandstones. The succession is probably 1200 m thick and includes in its upper part much quasi-brecciated material, once regarded as desiccation breccia but probably a product of diagenesis. Chert occurs only in beds of D_2 age.

In the Armagh area the basal clastics are overlain by pale limestones similar to the Dungannon succession and apparently of mid-Viséan age (S–D_1).

The Strangford outlier is no longer exposed but from the old quarries at Castle Espie, in pink crystalline limestones with a few shale partings, an extensive fauna of D_2 age was collected. Limestone blocks are common on many of the islands in Strangford Lough and the sub-drift outcrop may be quite extensive.

Upper Clastic Group

In the western areas where the top of the limestone succession is seen there is an abrupt change from calcareous sediments to shales or sandstones.

EXPLANATION OF PLATE 4

Some characteristic Lower Carboniferous corals

1; *Lithostrotion martini* Milne Edwards and Haime, a xl, b x2. **2**; *Lithostrotion pauciradiale* (McCoy), a x1, b x2. **3**; *Lithostrotion portlocki* (Bronn), a x1, b x2. **4**; *Lithostrotion aranea* (McCoy), a x1, b x2. **5**; *Lithostrotion junceum* (Fleming), a x1, b x2. **6**; *Amplexizaphrentis enniskilleni enniskilleni* (Milne Edwards and Haime), x1. **7**; *Caninia benburbensis* Lewis, x1.

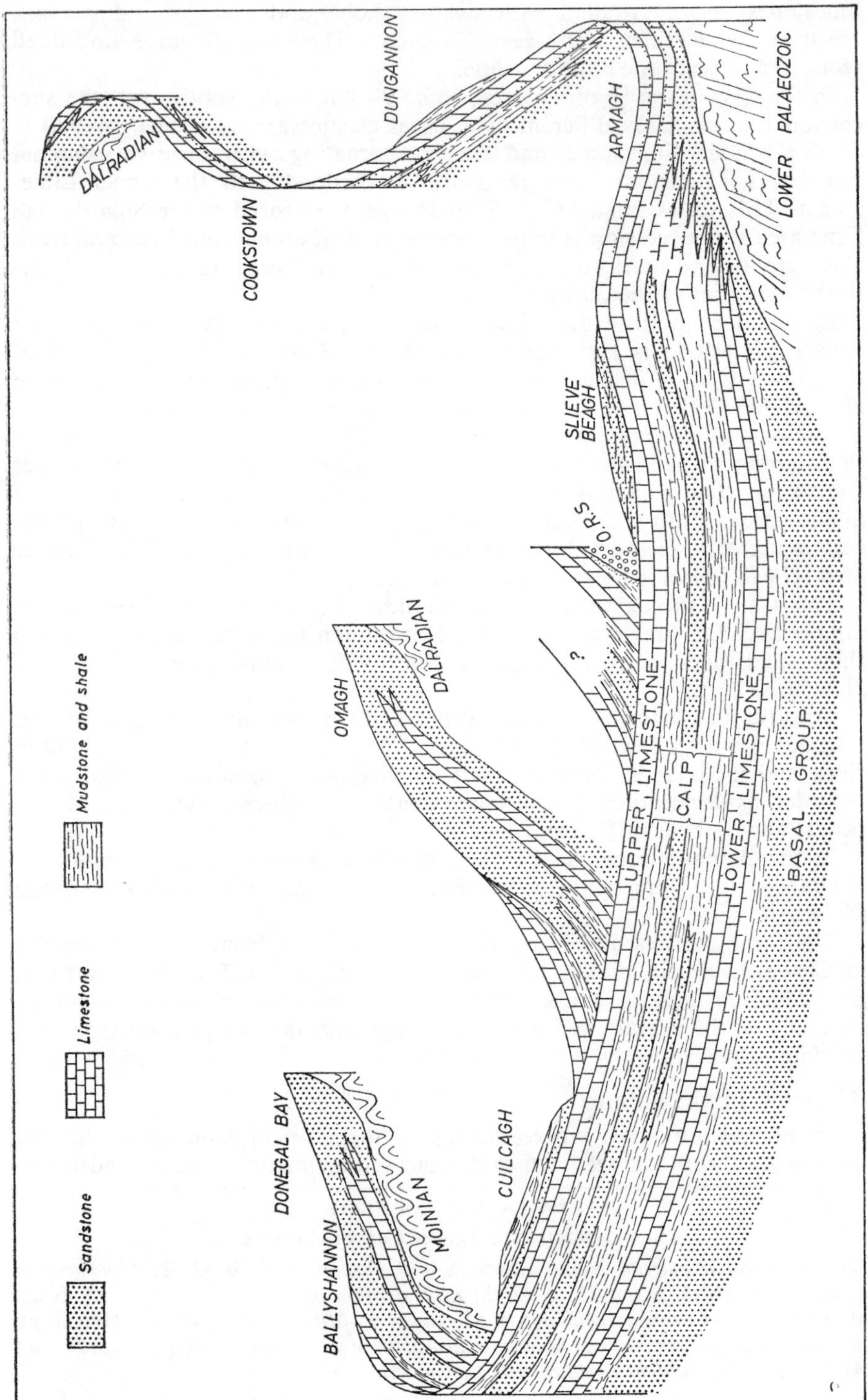

FIG. 15. *Ribbon diagram showing variation in Carboniferous lithology across Ulster*

Though without marked unconformity it is clear that there was a break in deposition and that clear water marine conditions gave place abruptly to coastal conditions.

In the Sligo syncline the Dartry Limestone on Benbulbin and the Dartry Mountains was succeeded by thin dark shales with impure limestone ribs—the Glenade Shales. This group is 60 m thick and is followed by the Glenade Sandstones—feldspathic sandstones of which some 60 m are preserved. The shales thin out to the east and the Kilroost reef limestones are overlain by the sandstone. Across the metamorphic anticline the limestones are overlain directly by the sandstone which forms an arcuate outcrop round Dough Hill and crops out over an area of over 130 sq km on the plateau north of Lough Macnean Upper, and over a somewhat smaller area to the south as far as Lough Allen and round Cuilcagh Mountain. The sandstones are generally massive, often coarse or conglomeratic, and quartzose. They are probably 120 m thick. An outlier forms the high ground of Slieve Rushen.

The sandstones are overlain in the areas from Kiltyclogher to Lough Allen and Lough Allen to Cuilcagh by a second group of shales, 210 m thick on Cuilcagh. A small outlier of these shales overlies the sandstones near Lough Fadd, west of Derrygonnelly. Soft and calcareous at the base, they become harder towards the top and contain bands of ironstone nodules. The shale on Dartry Mountain is of B_2 age. The lower half of the Cuilcagh shales, is of P_1 and P_2 age but the upper part of the Cuilcagh section is of E_2 age and will be considered in the next section (Namurian).

South of the Clogher Valley in the upland area centred on Slieve Beagh the limestones are overlain by massive sandstones, occasionally with coal streaks and dark shales with ironstone nodules, which are interbedded increasingly to the east with goniatite-bearing shales. In parts, at least, these beds show rhythmic banding but there are few calcareous beds and the succession is not a true 'Yoredale' one, as it is described in early memoirs. Goniatites in the lowest 100 m or so of the succession indicate ages from B_2 to P_1 and goniatite-bearing shales are exposed in the Fury River [H 567 487]. Thin red beds occur locally at several horizons. The total thickness is probably about 300 m.

In the Dungannon area the limestones are succeeded by the 180 m thick Rossmore Mudstones—a succession of shales and mudstones with a thin limestone near the base. The age of this group ranges from low P_2 to the P_2E_1 junction and the upper part of it is of Namurian age. As the underlying limestones are of D_2 age it is probable that the disconformity at the top of the limestones was later in this area than farther west.

Namurian

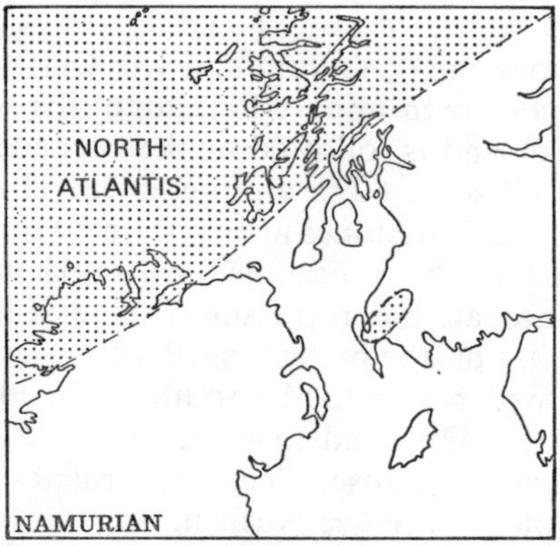

In the country round Lough Allen, in Co. Leitrim, the upper half of the so-called 'Yoredale' beds is now known to be of Namurian (E_2) age. These hard shales, containing crushed brachiopods, goniatites, and clayband ironstones, become somewhat sandy at the top and are overlain by the Millstone Grit which forms outliers on Cuilcagh, round Cullentragh Lough west of Lough Macnean, on the high ground north of Lough Allen, and on the Slieve Anierin hills to the east. The grit series, some 60 m thick, contains shale bands and two coal seams, one of workable thickness. It is overlain by 230 m of shales, originally classified as Lower Coal Measures but now thought to be of E_2 age.

In the Dungannon area (Fig. 16) the Rossmore Mudstones are succeeded by a series of grits, mudstones, and shales, with some coals and fireclays and one thin bed of dolomitic limestone, the Lurgaboy Dolomite. The occurrence of a dolomite of similar thickness and age in a borehole at Magilligan indicates that this horizon extended over a wide area. Exposures of these rocks are almost non-existent and our knowledge of them comes almost entirely from borehole records and the accounts of old pits and shafts which worked the coals. The base of the 'Millstone Grit' is taken at the bottom of the lowest grit, though the base of the Namurian is in the upper part of the Rossmore Mudstones, and the whole succession is about 550 m thick. Though some of the shales are very fossiliferous only two diagnostic specimens are known—*Tylonautilus nodiferus* 'early form' from some 60 m above the base of the succession is of E_1 age while *Hesperiella loudoni* from 160 m higher is of E_2 age.

Unlike the outcrops of the Namurian in west Ulster and Connaught, which are all relict outliers on hill tops, the Tyrone outcrop is a down-faulted block with a triangular outline running north-east from a base near Dungannon. It contained several workable coals the lowest of which, the Main Coal, occurs about 36 m above the base. Though the Main Coal was up to 1·5 m thick at outcrop it thinned rapidly to the east and has been worked out from shallow pits in the Drumglass area. Some 60 m above the Main Coal is a thin seam, the Crow Coal, and about 70 m above this the equally poor Congo Coal. Right at the top of the succession a series of workable coals occurs—the Sixteen-inch, Lower Two-foot, and Yard Coals. In general the succession is more arenaceous in its upper part.

No Namurian faunas younger than E_2 are known from the north of Ireland though Westphalian beds occur. This suggests that either there was a hiatus in sedimentation with a non-sequence covering H and R times or else there is a condensed sequence associated with non-diagnostic fossils. In Tyrone the mainly arenaceous sequence in the upper part of the Namurian

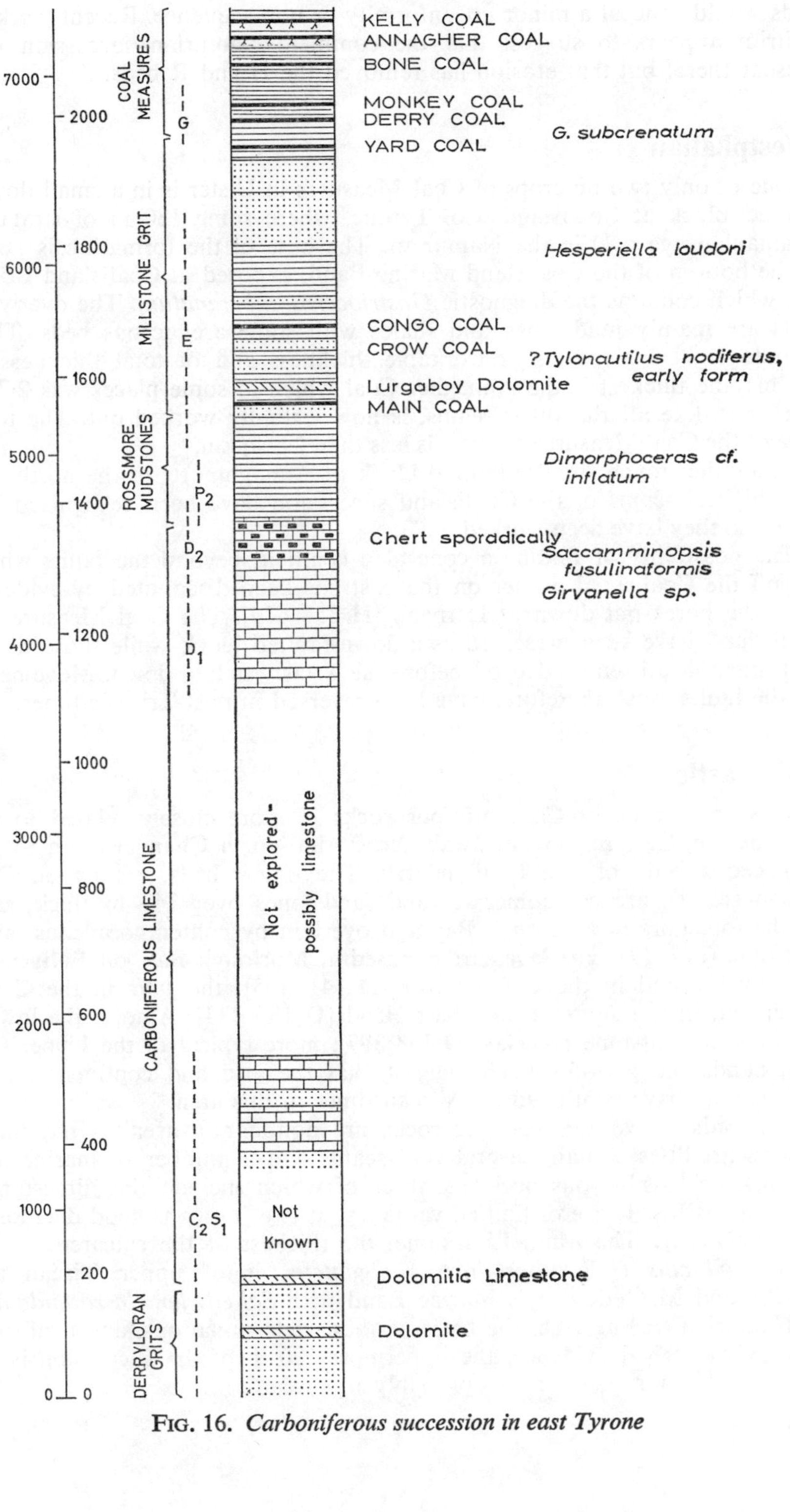

FIG. 16. *Carboniferous succession in east Tyrone*

beds would conceal a minor unconformity or non-sequence. Recent work in Leitrim appears to suggest that the complete Namurian succession was present there, but that erosion has removed the H and R beds.

Westphalian

One of only two outcrops of Coal Measures in Ulster is in a small down-faulted block at Coalisland, Co. Tyrone, where some 180 m of strata of Ammanian age overlie the Namurian. The base of the formation is taken at the bottom of the Coalisland Marine Band, exposed at Coalisland Brick-pit, which contains the diagnostic *Gastrioceras subcrenatum.* The overlying beds are mainly mudstones and shales with some arenaceous beds. They include eight coal seams of workable thickness and of total thickness of 9·7 m: the thickest is the Annagher Coal which at some places was 2·7 m thick but, like all the other seams, is now virtually worked out. The total area of the Coal Measures outcrop is less than 2·5 sq km.

In another much smaller faulted block at Annaghmore, to the north, the four highest seams of the Coalisland succession have been recognized but here also they have been worked out.

The possibility of finding a concealed coalfield beyond the faults which cut off the Coalisland outlier on the east has been discounted by evidence from the bores put down at Dernagh [H 854 670]. The Coal Measures at Coalisland have been preserved as a down-faulted block while those to the east must have been eroded off before the Trias was laid down. Movements on the faults must, therefore, have been reversed in post-Triassic times.

Ballycastle

This small area of Carboniferous rocks is more closely related to the sequence in Kintyre, 35 km away across the North Channel, than to the arenaceous beds of Co. Londonderry. The oldest beds, resting on Dalradian schists, are conglomerates and sandstones over 240 m thick, seen on the foreshore at Murlough Bay and overlain by contemporaneous lavas and tuffs (Fig. 17). The lavas are exposed at Murlough Bay, on Ballycastle golf course, and in the Carey River [D 141 405]; the tuffs in the Carey River and in the slopes below Fair Head [D 186 434]. Among the lowest beds are cementstone breccias [D 158 397], more typical of the Upper Old Red Sandstone period, which suggest that the arid and continental conditions may have continued till Viséan times in this area.

The beds above the volcanic rocks are dominantly arenites but interbedded argillites contain several coal seams and a number of marine and non-marine fossiliferous horizons, three of which include thin limestones. Most of the fossils are of limited value but at two horizons good diagnostic markers occur. The Main Limestone, the thickest of the calcareous beds with *Productus (Gigantoproductus) giganteus,* is of upper Viséan age (P_1–P_2), and McGildowney's Marine Band with *Schellwienella rotundata* is of Namurian (E_2) age. On the basis of these determinations, and confirmatory palynological evidence, the uppermost 100 m of the succession is regarded as Namurian, the rest is probably all Viséan.

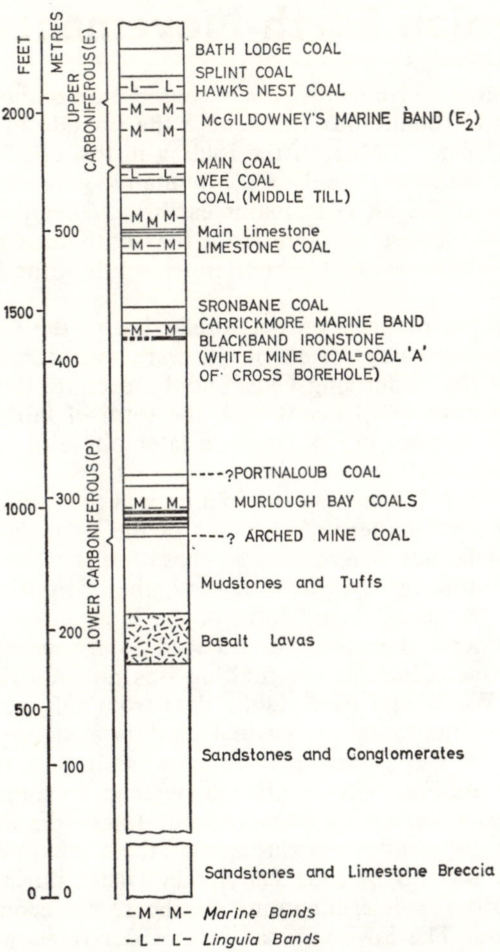

FIG. 17. *Carboniferous succession in the Ballycastle area*

The Namurian beds, and the Viséan succession above the volcanic rocks, are well exposed on the coastal cliffs between Ballycastle Bay and Fair Head, though the coal seams are never now seen at outcrop. The Carrickmore Barine Band can be examined at Carrickmore [D 165 427]; the Main Limestone at North Star [D 147 420]; McGildowney's Marine Band at Pollard [D 156 420] and Doon [D 141 416].

References

Adamson and Wilson 1933; Caldwell and Charlesworth 1962; Fowler 1955a; Fowler and Robbie 1961; George 1953, 1958, 1960; George and Oswald 1957; Hartley 1936a; Jackson 1965; Oswald 1955; Padget 1951, 1952, 1953; Robbie 1955b; Simpson 1955; Turner 1952; Wilson and Robbie 1966; Wright 1919, 1924a; Yates 1956.

7. Hercynian Earth Movements

The minor earth movements which caused non-sequences and unconformities during Carboniferous times were the prelude to a major orogeny at the end of the period. Most of the folding in this epoch was in the area now continental Europe—the Hartz Mountains gave their name to the period—and in the British Isles major earth movements were confined to the southern areas where pressure from the south caused intense folding of the Old Red Sandstone and Carboniferous, south of the Lower Palaeozoic foreland.

The effects of the orogeny on the foreland, and the Carboniferous and Old Red Sandstone which rested on it, were muted and folding, which tended to reflect the underlying Caledonoid structure, died out rapidly to the North. Over most of Ulster it took the form of fairly gentle folds on a N.E.–S.W. axis, though in Co. Cavan a later phase of E.–W. folding has also been detected.

More significant in Ulster than the folding was the faulting which accompanied it. The general pattern of outcrops of the Palaeozoic rocks in north Ireland has been largely determined by these faults which account for the Carboniferous outlier at Newtownstewart; the Namurian succession at Ballycastle; the preservation and form of the Dungannon–Coalisland coalfield; and the general shape of the Old Red Sandstone and Dalradian inliers in Co. Tyrone. Much of the faulting was on pre-existing Caledonian lines, but some E.–W. and N.–S. faults date from this period.

Though most of the faults are normal there is evidence of lateral movement on some, e.g. the Tempo–Sixmilecross Fault, and the Pettigo, Cool, and Castle Archdale Faults in the Omagh syncline. In some cases there was a renewal of movement on Caledonian fault lines—the Southern Uplands Fault was never quite quiescent throughout the Carboniferous.

Though there was no igneous activity in Ulster during the Hercynian period radiometric age determinations ascribe some secondary mineralization to this period. The low-temperature ores deposited were mainly lead, silver, zinc in the Silurian rocks of Co. Armagh and Co. Down, on N.–S. fault lines, and copper in the Old Red Sandstone of Tyrone. None of the deposits so far located are of commercial importance, with the exception of the Conlig lode, near Newtownards, which was extensively worked in the nineteenth century. Baryte was widely distributed as vein material and also occurs as disseminated intergranular cement in the Old Red Sandstone in Co. Tyrone. Baryte also occurs in large veins in the Carboniferous Limestone at Benbulbin, Co. Sligo.

8. Permo-Trias

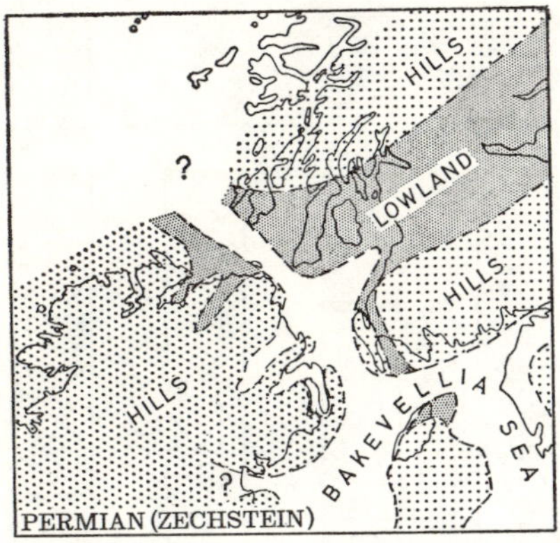

The first post-Armorican rocks were breccias and sandstones formed under arid conditions in faulted graben and resting on the eroded surface of Carboniferous and older rocks. On the low desert plains thin sands were locally blown into dunes. In Ulster such deposits are known only from the Lagan–Strangford area and the Lough Neagh basin where they have been found in boreholes. They are seen at one outcrop, the foreshore at Cultra —pronounced locally as 'Cultraw'— [J 412 809] where 1·5 m of brockram rest on Lower Carboniferous rocks. A thickness of 300 m of brockram and sandstone was found in a borehole near Scrabo [J 462 748], resting on Silurian greywacke, and 44 m (unbottomed) at Avoniel [J 361 739]. In Tyrone sands, sometimes with pebbles, from 1 to 8 m thick, are known from boreholes in the Grange area, near Stewartstown.

In late Permian times an arm of the "Bakevellia Sea" covered part of eastern Ireland. In the shallow and saline environment of the Bakevellia Sea and the contemporaneous Zechstein Sea beds of limestone, dolomite and a variety of evaporites were laid down in north Germany, the North Sea, and northern England. In Ulster the first marine deposit was the dolomitic Magnesian Limestone; 8·5 m thick at Cultra where it contains the Lower Zechstein fossil *Horridonia horrida;* 21 m at Avoniel; and 18 to 23 m thick in Tyrone, where it crops out in the east bank of the Killymoon River [H 823 750] and at Tullyconnel [842 749], whence it was first described by McCoy a century ago. In Tyrone the fossils found include *Batostomella columnaris, Thamniscus dubius, Bakevellia (Bakevellia) binneyi, Permophorus costatus, Schizodus obscurus, Glyptospira? helicina* and *cf. Naticopsis minima.* The Magnesian Limestone can probably be correlated with the Lower Magnesian Limestone of north-east England.

The limestone is succeeded by 1·5 m of gypseous mudstone and 1 m of grey dolomite—the Avoniel Limestone—which is correlated with the Fleswick Dolomite of west Cumberland. This passes up into 4 m of massive anhydrite and then a series of marls with thin beds of gypsum. These beds are over 70 m thick and though unfossiliferous are regarded as Permian.

The Highland Border Ridge, running south-westwards from Torr Head towards the Sperrins, appears to have been a partial barrier to sedimentation from the Palaeozoic onwards, but sandstones with rounded wind-blown grains of quartz, found at the bottom of the red sandstone sequence in the Port More Borehole [D 068 435], are tentatively ascribed to the Permian.

About 48 km south of the main outcrop an outlier of Permo-Trias at Kingscourt in Co. Cavan is of economic importance. A basal breccia up to 15 m thick rests on the Carboniferous and is overlain by 9 m of grey shale,

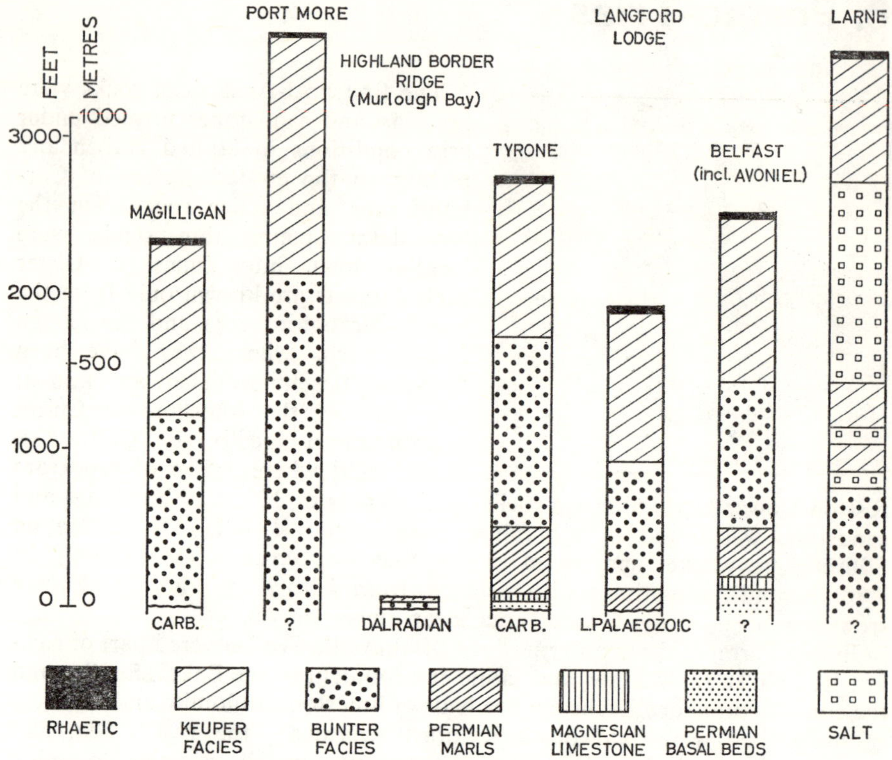

FIG. 18. *Comparative sections in the Permo-Triassic rocks*

the equivalent of the Hilton Plant Beds of Cumberland. The Magnesian Limestone is represented by 21 m of anhydrite and gypsum, which is separated by 9 m of marl from an upper bed of gypsum, 9 m thick. Succeeding red marls are also probably Permian in age and pass upwards into Bunter-type sandstones.

The Permian marls are succeeded by sandstones of Bunter facies which overstepped them on to the older rocks to the north and, probably, the west. This predominantly sandy formation is about 300 m thick in the Lagan Valley and Co. Tyrone, thins to almost nothing across the Highland Border Ridge and reaches a thickness of over 400 m in the Ballycastle–Ballymoney basin to the north where its base has not yet been found.

The Bunter sandstones are usually reddish brown in colour and were laid down in shallow water as indicated by mudcracks, ripple-marks, clay galls and very rare reptilian *(Chirotherium lomasi)* footprints. Borehole evidence shows that the succession contains very frequent mudstone partings, usually only a metre or so thick, and up to 30 per cent of the whole is argillaceous in character, though this is rarely apparent from surface exposures.

These beds are well exposed in old quarries at Scrabo [J 479 725] and in coastal sections at Red Bay. The breccias and sandstones of Armagh are also probably of this age.

At Murlough Bay, on the side of the Highland Border Ridge, a thin sequence of Triassic rocks is exposed in the cliff, with basal conglomerate, sandstones and marls all represented in a thickness of about 26 m.

The next major group of the Permo-Trias is of Keuper Marl Facies, consisting of soft silty mudstones of variable calcareous content, usually red but locally mottled with green and with subordinate bands—skerries—of grey and buff coloured sandstones, frequently dolomitic, which can be correlated over considerable distances. The marls display mudcracks and ripple-marks and also, locally, pseudomorphs of halite crystals. Though normally lacking in macro-fossils, these beds have, in the Dungannon area, yielded rare specimens of the small crustacean *Euestheria minuta* which may have lived in brackish water.

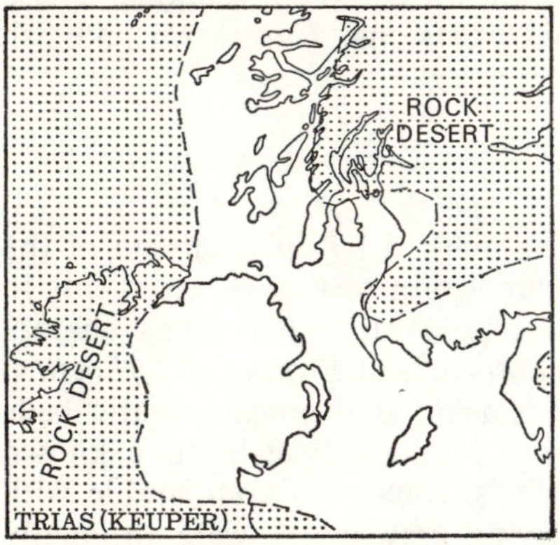

In south Antrim and Tyrone the Keuper is about 300 m thick, but north of the Highland Border Ridge a thickness of 670 m has been recorded in the Port More Borehole [D 068 435] where basement beds and a Keuper Waterstone group have been recognized. In the Larne area a considerable thickening is due to the occurrence of a massive development of rock salt. The salt field extends north-eastwards from the Carrickfergus area, where the salt has a total thickness of 40 m, with individual beds up to 27 m, to Larne, where it totals 488 m in the main groups. The mode of formation of these great quantities of salt is now thought to be by crystallization from highly saline waters trapped in localized down-warps on the floor of a shallow shelf sea on the continental margin (Fig. 19).

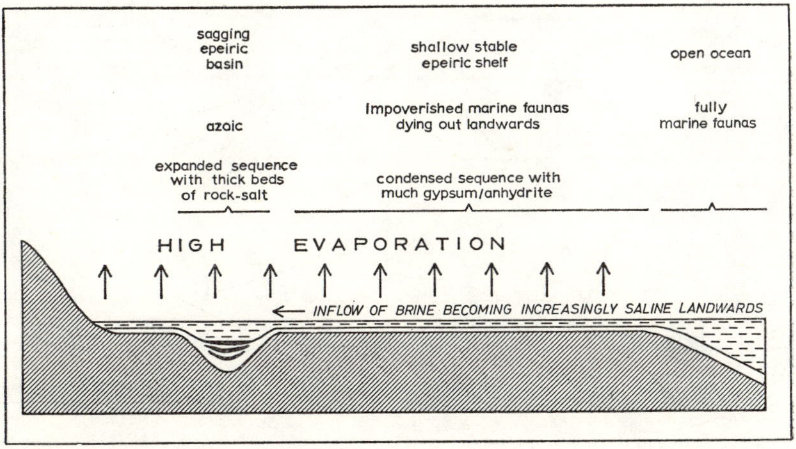

FIG. 19. *Postulated depositional environment of the Triassic salt beds*

The present outcrop of the Keuper is largely on the north side of the Lagan Valley but beds of this unit are seen in coastal exposures as far north as Glenariff and in counties Tyrone and Londonderry. Good exposures are seen at present in brick pits at Collin Glen [J 272 715] and Forth River [J 291 770].

The upper part of the Permo-Trias has yielded a microflora of plant spores and pollen which shows that the Keuper facies of County Antrim is equivalent to the Keuper Sandstone, Waterstones and Keuper Marl of Worcestershire and to the Upper Bunter, Muschelkalk, and Keuper divisions of the German Trias.

The onset of deeper-water conditions at the end of the Triassic period is indicated by the development of the Tea Green Marls, some 10 m thick, at the top of the succession in all of the known complete sections. These pale greyish green silty mudstones contain occasional *Euestheria* and fish debris.

Rhaetic

The last stage of the Triassic period was marked by an incursion of the sea—the Rhaetic transgression—and the deposition of marine sediments in the form of dark shaly mudstones with an abundant bivalve fauna characterized by the presence of *Rhaetavicula contorta* and *Protocardia rhaetica*. Towards the base of the series there are several thin conglomeratic beds composed almost entirely of organic debris, particularly fish teeth, known as 'bone beds'. Higher beds include red marls, somewhat comparable with marls in the Keuper. The red and overlying grey marls were originally and erroneously called the 'White Lias'.

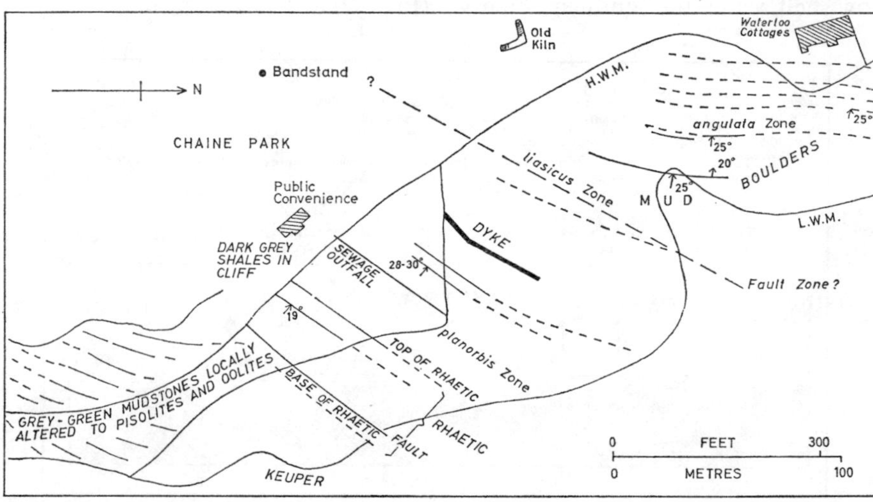

FIG. 20. *Sketch-map of the Rhaetic and Lias exposures on the foreshore at Waterloo, Larne*

The Rhaetic beds in Ulster, though usually less than 15 m thick, are widely preserved beneath the overlying Jurassic sediments and can be seen at outcrop at a number of localities around the Antrim plateau. The classic locality at Collin Glen is now obscured by bridge works and the best exposures are at Waterloo, Larne, where foreshore outcrops [D 408 037] show 12·5 m of grey shale and mudstones with thin limestone and sandstone bands and 'bone beds' (Fig. 20). The cliff section here is affected by local tectonic and igneous activity and the full thickness is not seen. In the Larne Borehole a total of 20 m of dark shales are ascribed to the Rhaetic. Rhaetic beds are well exposed in a landslip just west of Whitehead.

The Rhaetic is probably present below the Lias, wherever it occurs, for 20 km or so north of Larne but is rarely seen because of the extensive landslips. To the west it is known only from boreholes—at Langford Lodge it is 17 m thick and in Co. Tyrone 10 m of Rhaetic, with '*Schizodus ewaldi*' and *Gyrolepis,* were found at Mire House [H 861 663].

North of the Highland Border Ridge Rhaetic beds are known from boreholes at Magilligan (22 m) and Port More (4 m but partly assimilated by an intrusion) and are exposed in a stream section in The Linn [C 709 279] where dark shales rich in *Rhaetavicula contorta* are seen.

References

Ashley 1946; Evans 1968; Fowler 1955b; Fowler and Robbie 1961; Hartley 1943, 1949; King 1853, 1857; Manning, Robbie and Wilson 1970; Reynolds 1928; Sherlock 1926, 1928; Stubblefield 1949.

9. Jurassic

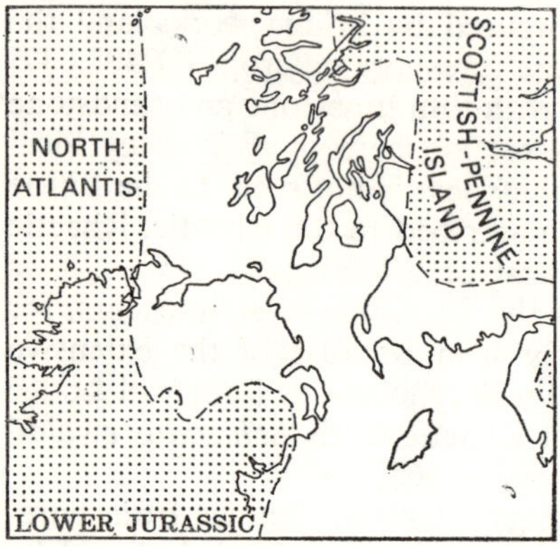

The marine deposits of the Rhaetic were succeeded by a substantial thickness of Liassic mudstones and limestones laid down in conformity with the underlying beds. It is possible that, as in other parts of the British Isles, the area of Liassic deposition was greater than that of the Rhaetic. No instances are known, however, in Ulster of Lias over-stepping the Rhaetic.

Like the Rhaetic, Lower Lias rocks crop out around the periphery of the Antrim plateau and have also been proved in boreholes west of this area. The Lias is dominated by mudstones with subordinate limestone developments, particularly in the *angulata* and *bucklandi* Zones (Fig. 21). The soft grey mudstones are commonly calcareous and also contain siltstones, limestone ribs and occasional ironstone nodules. These incompetent beds, lying beneath the massive Cretaceous White Limestone, are very liable to cause landslips so that the Lias is only occasionally seen in an undisturbed state. Surface outcrops rarely exceed 30 m in thickness though substantially greater thicknesses are known from deep boreholes.

At the base of the Lias, the pre-*planorbis* beds of the *planorbis* Zone are known from Waterloo (Larne) [D 409 037] where they are seen to overlie the Rhaetic conformably (Fig. 20); they have been reported from White House, Island Magee [J 457 985], and Collin Glen [J 269 722]. The Waterloo

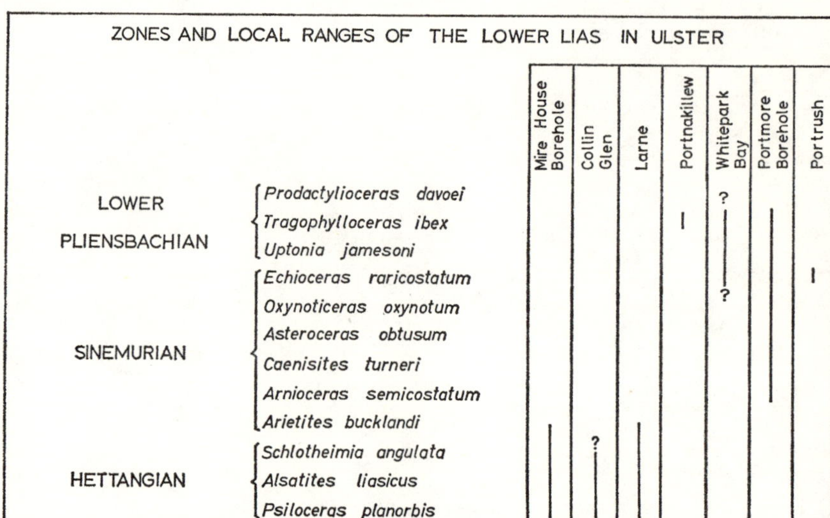

FIG. 21. *Zones and local ranges of the Lower Lias in Ulster*

PLATE 5

Typical Rhaetic and Lower Lias fossils

section shows continuous exposures well into the *planorbis* Zone and intermittently above. In the Larne Borehole over 30 m of beds belong to this zone. Fissile mudstones with *Psiloceras planorbis* and *Liostrea hisingeri* are frequently found in the slipped Lias alongside the coast road as far north as Straidkilly [D 305 163] and also at White House. The zone is also known in Derry from a deep borehole at Magilligan [C 683 353] where a total of 90 m of Lower Lias was penetrated below a massive sill. The higher zones of the Hettangian are best known from scattered exposures in south and east Antrim. The exposures currently visible in Collin Glen (about 5 m) belong to the *liasicus* and possibly the lowest part of the *angulata* zones. The Mire House Borehole [H 861 664] proved about 90 m of beds attributed to the Hettangian. The *bucklandi* Zone of the Lower Sinemurian is also known to be present in the Larne area (Fig. 21).

In north Antrim higher Liassic rocks are exposed at Portrush, in White Park Bay (Ballintoy) and at Portnakillew. At Portrush they have been baked by the Portrush Sill to give a tough hornfels which looks like an igneous rock but contains recognizable fossils including ammonites. This 'Portrush Rock', which is well displayed on the foreshore east of Lansdowne Terrace [C 857 410], was thought by the Neptunists in the early days of geology to prove that igneous rocks were formed by sedimentary processes. The ammonites present include *Paltechioceras* indicating the *raricostatum* Zone of the Upper Sinemurian.

At White Park Bay exposures vary with the state of the sand beach and the blown sand dunes behind it. Sections at Oweynamuck [D 029 448] prove the occurrence of the *raricostatum* and *ibex* Zones in grey mudstones; temporary sections in the beach within the bay also reveal *raricostatum* Zone mudstones. The fauna includes the ammonites *Cheltonia, Hemimicroceras, Gemmellaroceras, Oxynoticeras* and *Paltechioceras* together with the gastropods *Procerithium* and *Zygopleura* and the bivalves *Camptonectes, Cardinia, Chlamys, Entolium, Grammatodon, Liostrea, Modiolus, Nuculana, Oxytoma, Pleuromya, Plicatula, Protocardia* and *Pseudolimea.* Though zones as high as *davoei* and as low as *planorbis* have been claimed to occur at White Park these other zones are not now exposed and their presence seems improbable in view of the sequence found in the Port More Borehole, only 4·8 km to the east.

EXPLANATION OF PLATE 5

Typical Rhaetic and Lower Lias fossils

1. *Rhaetavicula contorta* (Portlock), cast of holotype, Rhaetic, Westbury Beds. **2.** *Psiloceras planorbis* (J. de C. Sowerby) *planorbis* Zone and Subzone, side view; **3.** *Psiloceras* (*Caloceras*) *intermedium* (Portlock), *planorbis* Zone, *johnstoni* Subzone, side view and venter; **4.** *Plagiostoma gigantea* J. Sowerby; **5.** *Gryphaea arcuata* Lamarck; **6.** *Schlotheimia* (*Waehneroceras*) *prometheus* (Reynès), *liasicus* Zone, *portlocki* Subzone, side view and venter; **7.** *Leptechioceras macdonnelli* (Portlock) *raricostatum* Zone, *macdonnelli* Subzone, side view and venter.

All natural size except Fig. 1 which is x2; Fig. 7 reproduced by kind permission of the Trustees of the British Museum (Natural History).

The small exposure at Portnakillew [D 090 434] yielded a fauna of *ibex* Zone age whilst Liassic mudstones caught up in the Tertiary volcanic vent of Carrickarade are believed to be of *jamesoni* Zone age.

The deep borehole at Port More [D 068 435] penetrated 269 m of Lower Lias above a complex sill which may have assimilated lower beds. It proved a sequence from the *ibex* Zone down to the *semicostatum* Zone.

No Lias of more recent age is known at outcrop in Ulster, but derived fossils of Middle and Upper Lias age have been recorded from the basal Cretaceous conglomerates in north Antrim. Blocks of shale with Middle Lias fossils have been found in the drift in the Ballycastle district and it is probable that beds of this age outcrop below the North Channel or below sea level around Rathlin Island.

There is no evidence that any deposits of Jurassic age more recent than the Lias were laid down in the Ulster area, but as Middle and Upper Jurassic beds as high as Kimmeridgian in age are known in Skye and Mull it is not impossible that more recent Jurassic beds were once present in north-east Ireland but were lost during the pre-Cretaceous erosion.

References

Fowler and Robbie 1961; Langtry 1874; Tate 1864, 1868, 1870; Wilson and Robbie 1966.

PLATE 6

Typical Cretaceous fossils

10. Cretaceous

There are no representatives in Ulster of the Lower Cretaceous rocks and there was clearly a long period of non-deposition in the early part of the Cretaceous era during which much of the upper part of the Jurassic system may have been removed by erosion. The occurrence of Chalk in a collapse breccia in Co. Kerry, and in outliers on the Continental Shelf to the west of Ireland suggests that the Cretaceous beds were deposited over the whole Irish area, although today they are found only in Ulster where they have been protected from erosion by the capping of Tertiary basalts. Cretaceous rocks underlie much of Co. Antrim and eastern Londonderry, appearing at outcrop around the edge of the lava plateau and in a few inliers. West of Lough Neagh a small outlier is found on Slieve Gallion, a considerable thickness has been proved in the Aughrimderg Borehole near Dungannon, and there is a limited outcrop below the lavas north of Coalisland.

The Cretaceous sediments of Ulster can be conveniently divided into a lower arenaceous sequence—the Hibernian Greensand—which includes strata of various ages up to and including the *Micraster coranguinum* Zone; and an upper calcareous sequence known as the White Limestone or Chalk, which ranges up to the Lower Maestrichtian. Sedimentation was repeatedly interrupted by periods of uplift and erosion, so that the succession at any one place includes more or less extensive non-sequences. The sedimentary history is even more complex than is apparent from earlier literature, but it is now known in considerable detail as a result of the unpublished researches of T. P. Fletcher, which have been drawn on extensively in this chapter. Figure 22 shows the stratigraphy in a number of key areas, and draws attention both to the diachronous nature of the onset of White Limestone sedimentation and the extent to which the local succession was reduced through erosion prior to the extrusion of the Tertiary basalts. It is now known that the various components of the Hibernian Greensand have a much wider distribution than was formerly thought, and that their absence from other areas is due to intra-Cretaceous erosion rather than to non-deposition. It is also likely that White Limestone of Maestrichtian age was deposited over the whole area, although today it is preserved only in those areas which were protected from subsequent erosion by down-folding or faulting.

EXPLANATION OF PLATE 6

Typical Cretaceous fossils

1. *Schloenbachia subvarians* Spath; **2.** *Chlamys fissicosta* (Etheridge); **3.** *Turritella unicarinata* (Woodward); **4.** *Scaphites elegans* Tate; **5.** *Neoliothyrina obesa* Sahni; **6.** *Exogyra obliquata* (Pulteney) [synonym *E. conica* (J. Sowerby)]; **7.** *Exogyra columba* (Lamarck); **8.** *Entolium orbiculare* (J. Sowerby).

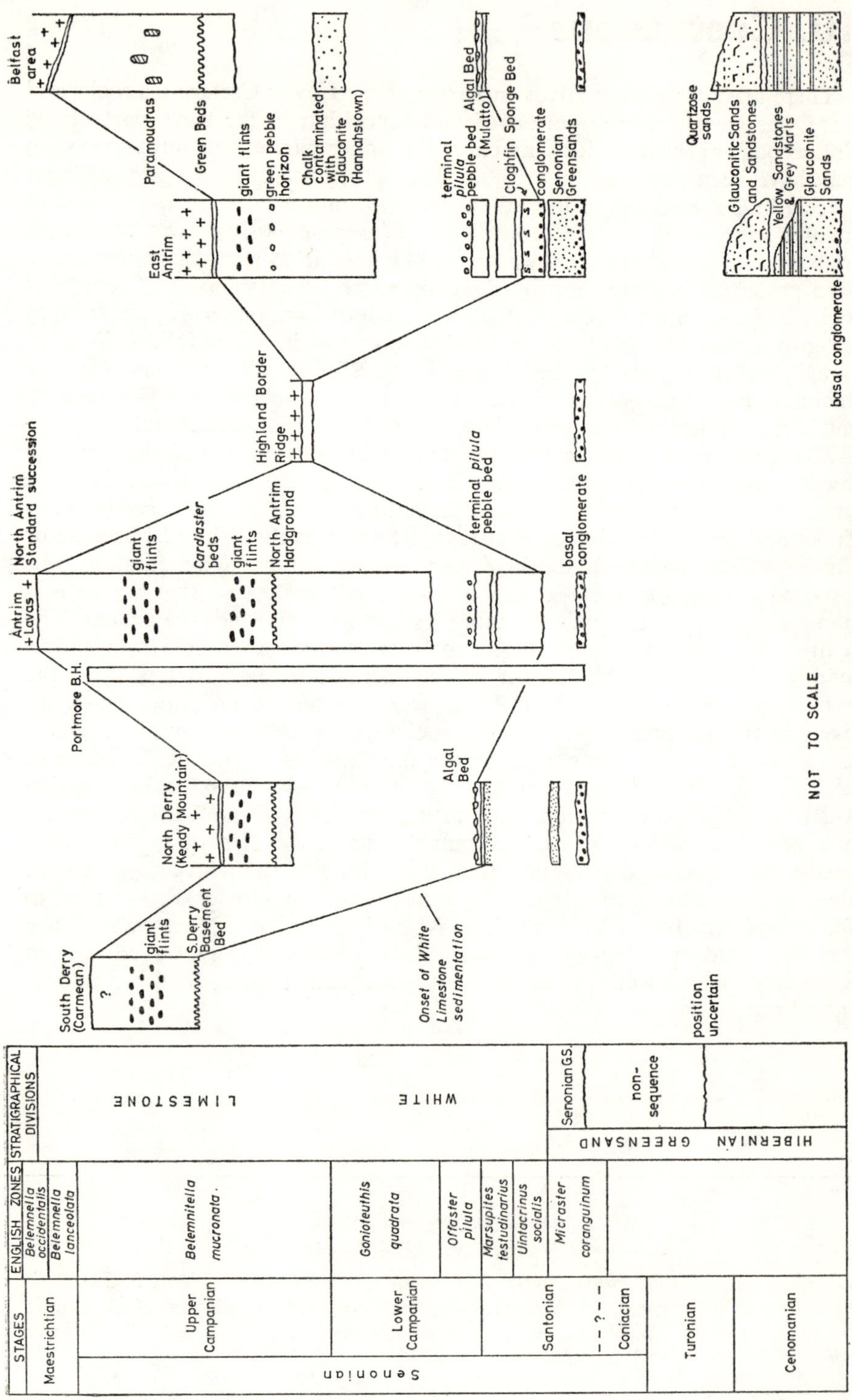

FIG. 22. *Zones and local ranges of the Cretaceous rocks in Ulster*

A. *White Limestone (Chalk) underlying basalt lavas. Magheramorne Quarry.*

PLATE 7

B. *The Tertiary granite mountains of the Western Mournes. From Bushtown.*

A. *Tertiary basalt lavas overlying White Limestone (Chalk) with flint bands. Rotational landslips. Garron Point.*

PLATE 8

B. *'The Lord'. A raised-beach stack of massive basalt lava standing on the vesicular top of the underlying flow. Island Magee.*

Hibernian Greensand

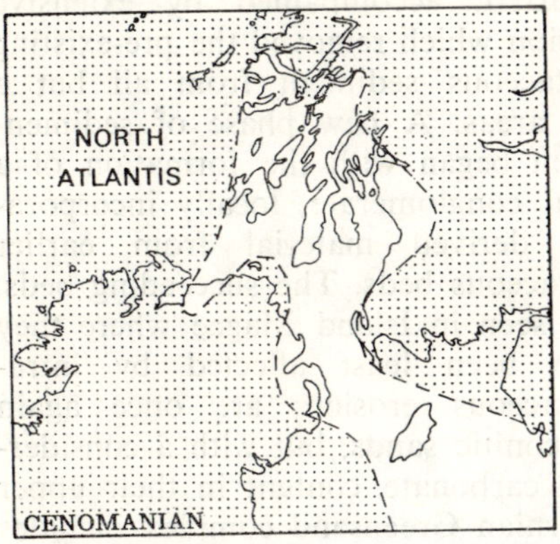

The Hibernian Greensand comprises a complex sequence of arenaceous sediments which includes one minor and one major non-sequence. A lower subdivision in which questionably Turonian sediments rest non-sequentially on sediments of Cenomanian age is overlain unconformably by Greensands of Senonian age. The Hibernian Greensand complex is preserved in the Belfast area and on the east coast, but elsewhere it has been largely if not completely removed by erosion.

The oldest beds are some 4 m of poorly consolidated dark green glauconitic calcarenites termed the Glauconite Sands. These are highly fossiliferous at certain levels and yield a rich molluscan fauna dominated by *Exogyra obliquata* (the *E. conica* of earlier literature) associated with *Entolium orbiculare, Chlamys* (*Aequipecten*) *aspera, C. fissicosta* and several species of *Oxytoma,* with Lower Cenomanian ammonites including *Mantelliceras* and *Schloenbachia.* The occurrence of the belemnite *Actinocamax primus* at certain localities is noteworthy. The basal sands are succeeded by the Yellow Sandstones and Grey Marls, calcareous buff-coloured silty sandstones alternating with grey marls. Irregular cherty concretions are present at some levels. This sequence reaches a maximum thickness of 9 m at Kilcoan, Island Magee [J 460 988], but elsewhere, e.g. Collin Glen [J 269 722] (7·3 m) and Woodburn Glen [J 383 896] (1·2 m), the thickness is reduced through intra-Cretaceous erosion. The fauna is characterised by *Arctostrea colubrina,Nithea spp.,* echinoids including *Hemiaster sp.,* and small rhynchonellids provisionally referred to *Lamellaerhynchia.* The occasional ammonites, which include species of *Acanthoceras, Austiniceras* and *Schloenbachia,* indicate a Middle Cenomanian age.

A well-marked erosion surface separates the Yellow Sandstones from the succeeding Glauconitic Sands and Sandstones, which are best seen at Collin Glen, but elsewhere have been largely if not completely removed by erosion. The lower beds, 4·5 m thick, are glauconitic, but pass upwards into 6 m of pure quartzose sands. Successive levels within the glauconitic beds are characterized by serpulids, large specimens of *Exogyra columba* occurring in shell-beds and bryozoa associated with appendages of the small crustacean '*Callianassa*'. One bed is literally packed with the small brachiopod '*Waldheimia hibernica*' which is probably a zeilleriid. Apart from an unconfirmed early record of '*Ammonites lewesiensis*'—suggesting a Turonian *Lewesiceras*—the age of these sands is doubtful, although by comparison with French successions the occurrence of *Exogyra columba* would suggest a high Cenomanian and/or low Turonian position. The Quartzose Sands, which can be compared with the Loch Aline (Argyllshire) Glass Sands of similar uncertain stratigraphical position, have so far yielded no macrofossils, while the evidence from the limited microfauna is equivocal.

A prolonged period of uplift then occurred, accompanied by extensive erosion which removed the pre-existing Cretaceous sediment from all but a few areas. A new phase of sedimentation began with the formation of a basal conglomerate, locally incorporating derived material from earlier Cretaceous beds. The succeeding beds, best seen in Island Magee where they have been least affected by intra-Cretaceous erosion, are once again glauconitic sands, but with a considerable carbonate content in their upper levels. This uppermost part of the Hibernian Greensand complex includes several laterally continuous bands packed with thick-shelled *Inoceramus* fragments and has yielded depressed forms of *Micraster; Conulus* and *Echinocorys* of *Micraster coranguinum* Zone type; and long-ranged bivalves such as *Spondylus spinosus* and *Chlamys* (*Aequipecten*) *aspera.* The brachiopod fauna comprises predominantly rhynchonellids below the *Inoceramus* bands, including *Cretirhynchia robusta* and an undescribed inflated group, while the higher beds yield terebratulids e.g. *Gibbithyris hibernica*, these being locally abundant in Island Magee. Although the echinoid fauna suggests a *Micraster coranguinum* Zone age for these Senonian Greensands the possibility cannot be excluded that beds of *M. cortestudinarium* Zone age are also represented. It is difficult to assess whether or not certain brachiopods and echinoids of *cortestudinarium* Zone aspect are merely facies forms continuing in arenaceous sediments belonging to the *coranguinum* Zone. Similar arenaceous deposits are exposed in streams around the flanks of the Binevenagh plateau in the Limavady area, and also in Tircrevan Burn [C 703 322], where they overlie a basal 'bone-bed' containing derived Jurassic material. Here they correspond to the *Uintacrinus, Marsupites* and *pilula* zones of the north Antrim coast, though the index crinoids have not been found.

A further period of uplift and erosion followed which removed the Senonian Greensands in all but a few areas (Fig. 23). A new period of marine transgression initiated the deposition of the White Limestone, which begins with a basal conglomerate incorporating residual material from the Senonian Greensands, including the basal conglomerate of that subdivision which itself incorporates material from the underlying Cretaceous beds. In this context the occurrence of a phosphatized *Lewesiceras mantelli* in the conglomerate at Torr Head is of great importance in that it clearly demonstrates the former existence of Turonian sediments.

White Limestone

Following the deposition of the arenaceous sediments of the Hibernian Greensand complex there was a marked change in the type of sedimentation, and thereafter only calcareous rocks were formed. The Chalk in

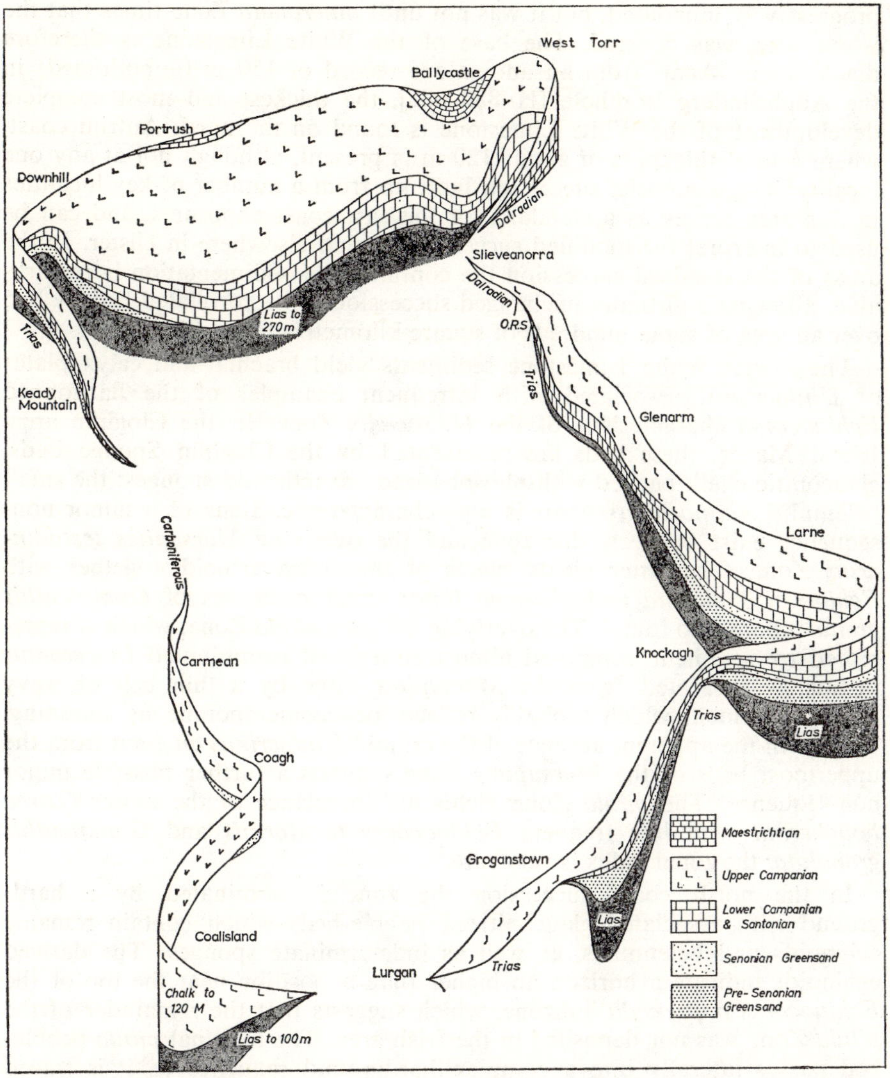

FIG. 23. *Ribbon diagram showing variations in the Lias and Cretaceous successions in Ulster. This is intended to give only a general impression of the stratigraphy and is not to scale. Faults are not shown.*

Northern Ireland differs from the contemporaneous English Upper Chalk in that it is exceptionally hard and impervious as the result of diagenetic recrystallization of calcite in the pore spaces of the original matrix. Stylolites and other pressure-solution phenomena are present throughout and the fossils are difficult to extract. Courses of flint, both nodular and tabular, occur at most levels.

Throughout the period of White Limestone deposition the upstanding massifs of older rocks such as the Dalradian Highland Border Ridge were progressively inundated, but it was not until *mucronata* Zone times that the whole area was covered. The base of the White Limestone is therefore diachronous. Apart from an anomalous record of 150 m (unbottomed) in the Aughrimderg Borehole [H 880 685], the thickest and most complete development of the White Limestone is found on the north Antrim coast, where a total thickness of about 120 m is present, although not at any one locality. The composite succession built up from a number of key localities in this area serves as a standard for the east coast exposures, and can be used to interpret the modified successions found elsewhere in Ulster. In the areas of the standard succession the continuity of sedimentation is remarkable, allowing a virtually unchanged succession to be correlated bed by bed over an area of some hundreds of square kilometres (Figs. 22, 23).

The lowest White Limestone sediments yield brachial and calyx plates of *Uintacrinus,* associated with infrequent examples of the flat-topped *Echinocorys* characteristic of the *U. socialis* Zone. In the Cloghfin area, Island Magee, these beds are represented by the Cloghfin Sponge Beds, glauconitic chalk packed with phosphatized hexactinellid sponges; the small belemnite *Actinocamax verus* is also characteristic. Signs of a minor non-sequence exist between this zone and the overlying *Marsupites testudinarius* Zone. The latter yields plates of the index crinoid, together with *Echinocorys elevata* and *Kingena lima;* small examples of *Gonioteuthis granulata* are also found. The overlying *Offaster pilula* Zone, which is represented by sediment composed almost entirely of comminuted *Inoceramus* prisms, is separated from the *Marsupites* Zone by a thin belt of wavy bedded sediment which probably reflects penecontemporaneous slumping. This, with the apparent absence of the crinoid *Uintacrinus anglicus* from the uppermost beds of the *Marsupites* Zone suggests a further possible minor non-sequence. The *pilula* Zone yields an abundance of the oyster *Ostrea boucheroni* with less frequent *Echinocorys tectiformis* and *Gonioteuthis granulata;* the zonal index is very rare.

In the north coast succession the zone is terminated by a hardground with associated glauconitized pebble-beds which contain remanié echinoids and belemnites, as well as indeterminate sponges. The derived echinoids indicate a horizon no higher than a position near the top of the *Echinocorys depressula* Subzone, which suggests that the remainder of the *pilula* Zone was not deposited in the Irish area. The terminal *pilula* pebble-bed passes laterally into a complex hardground sequence as it is traced away from the area of standard succession, and in its extreme development is capped by a layer of algal stromatolites (e.g. the Belfast Mulatto), indicating a very shallow depth of formation. In the Belfast area and Co. Derry (e.g. Keady Mountain) the well-developed algal bed is indicative of a greater period of non-sequence than elsewhere. Above the major non-sequence at the top of the *pilula* Zone a considerable thickness of chalk can be referred to the *Gonioteuthis quadrata* Zone and preliminary studies indicate that only the lowest subzone of the English succession is missing. *Echinocorys* typical of the *Saccocoma cretacea* Subzone in England occurs in a band near the base of the succession, with important marker horizons yielding respectively slim *Belemnitella praecursor* and *Orbirhynchia bella* a few metres higher. The *Orbirhynchia* band has also yielded a single calyx of the subzonal crinoid. The zonal index occurs throughout. The uppermost beds are characterized by the small subspecies *G. quadrata gracilis* as on the Continent, and show a gradual replacement of *Gonioteuthis* by large elongate belemnites belonging to the group of *Belemnitella praecursor*, and the incoming of a fauna foreshadowing that of the overlying *Belemnitella mucronata* Zone, e.g. *Echinocorys conica* and *Cretirhynchia woodwardi*.

Apart from one or more non-sequences the Antrim *mucronata* Zone represents the most complete and best exposed succession from the *quadrata* Zone to the Maestrichtian in north–west Europe. The standard European belemnite assemblage sequence *Belemnitella senior–B. minor–B. langei* can be recognized, and the succession of non-belemnite faunas broadly follows that found in the Norfolk Chalk. Excluding the basal beds, characterized by *Echinocorys conica, Cretirhynchia woodwardi,* small *Kingena pentangulata* and *Belemnitella senior,* the remainder of the zone can be divided into two parts, the line of division approximately coinciding with an important hardground which is developed on the north coast at about the level of the Catton Sponge Bed of Norfolk. In the east coast sections this hardground is represented by a horizon of green pebbles. The lower sequence includes beds characterized by *Galeola* and *Offaster* respectively. Within the higher sequence a further subdivision can be made. The lower beds comprise two belts with giant flints separated by beds with a rich fauna of *Cardiaster granulosus, Galerites* and *Cretirhynchia norvicensis.* Above this are beds with a fauna showing a progressive incoming of Maestrichtian elements such as thick-tested *Echinocorys* of the *belgica* group, *Cretirhynchia magna, C. retracta* and columnals of the crinoid *Austinocrinus bicoronatus.* A full succession from the north Antrim hardground into the Maestrichtian can be seen in the White Rocks, Portrush, and the adjoining Ballymagarry Quarry. The lower part of the *mucronata* Zone is best studied in the coast sections between Ballintoy Harbour and Larry Bane Bay, and in Capecastle quarry [D 101 365].

The Maestrichtian beds are preserved in the down-faulted block between the Portnakillew and Tow Valley faults, and also in a relatively downfolded area near Portrush. The most important sections are seen in Port Calliagh and thence eastwards to Ballycastle, where the highest beds are exposed near the harbour. The base of the stage is taken at the incoming of the belemnite *Belemnella,* there being no major non-sequence between the Campanian and Maestrichtian in Ireland as there is elsewhere in Europe. A lower *Belemnella lanceolata* Zone contains the index belemnite associated with late forms of *Belemnitella langei.* The remaining fauna includes *Echinocorys belgica, E. heberti, Chatwinothyris spp., Cretirhynchia magna* (common), *C. retracta, Magas chitoniformis* and *Terebratulina gracilis.* The higher zone is characterized by an abundance of the index belemnite *Belemnella occidentalis.* Derived Maestrichtian microfossils are present in the 'Ballycastle Pellet-Chalk', a reworked deposit of pre-Tertiary Basalt age.

When traced into the southern and western areas important modifications are seen in the standard *mucronata* succession. At Hannahstown near Collin Glen the basal *mucronata* White Limestone is locally heavily contaminated with glauconite derived from older Cretaceous sediments. In general the base of the White Limestone rises westwards, and over much of the area sedimentation began relatively late in *mucronata* times. A major hardground, developing at about the level of the *Cardiaster* beds of the north coast—i.e. considerably higher in the succession than the north Antrim hardground—extends laterally into the 'Green Bed' complex found near the base of the White Limestone in the Lisburn area, and forms the richly fossiliferous basement bed in southern Derry around Moneymore. The fauna is comparable in preservation with the Chalk Rock *reussianum* fauna of England, and includes a large number of ammonites, as well as bivalves and gastropods. The overlying beds, which contain giant flints on the north coast, are represented in the Belfast area by White Limestone with the giant vertical cylindrical flints known as paramoudras. The uppermost beds both in this area and in Derry probably do not quite reach the Maestrichtian, with the possible exception of White Rock near Belfast [J 295 742]. The enormous thickness recorded in the Aughrimderg borehole remains inexplicable: unless the succession has been repeated by faulting it may be that Maestrichtian sediments are present. Similarly, nothing is known of the White Limestone in the centre of the basalt plateau, apart from the small 'inlier' at Templepatrick. Borehole evidence has shown that the chalk here is underlain by a great thickness of intrusive rhyolite, so that it may well be a raft lifted by the intrusion.

References

Fletcher 1967; Hancock 1961; Hume 1897; McGugan 1957; Reid 1958, 1959, 1962, 1963, 1964, 1971; Tate 1865; Wilson and Robbie 1966; Wood 1967.

11. Tertiary

General Introduction

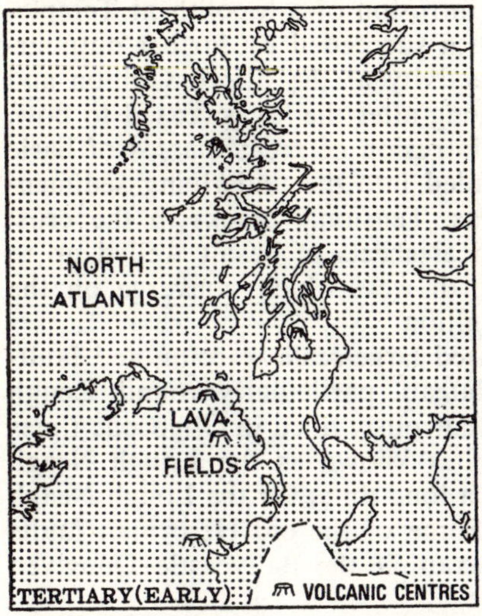

The end of the Cretaceous period saw the final retreat of the sea from the region of Ulster and the elevation of the Chalk to form a new land surface. The climate was apparently like that which had prevailed during the deposition of the Chalk—arid and warm—for the erosion of the Chalk seems to have been a slow and gentle process, the soluble limestone being gradually leached out to leave a 'soil' of flints and red lateritic clay. There is little evidence to be seen now of much deeper erosion and the clay and flint land surface was generally fairly flat and regular. Where it can be seen it consists of a bed of red clay and flints which averages 2 m thick, reaching a thickness exceptionally of about 6 m. There is some evidence of grike formation in the underlying chalk. The clay was presumably derived largely from the insoluble residue of the chalk while the quantity of flint present suggests that a moderate thickness of limestone disappeared during the early Tertiary. Traces of lignite indicate plant growth at least in places on the Palaeocene landscape.

This prolonged period of denudation and tectonic quiescence was ended abruptly in the Palaeocene or Eocene by the outbreak of widespread volcanic activity which continued for many millions of years in the British Isles and is still going on in Iceland. Remnants of the extrusive lavas and tuffs and intrusive plugs, sills, batholiths, dykes, etc. which were formed at this time are found in a wide arc stretching from Greenland through Jan Mayen Land, Iceland, the Faeroes and the Inner Hebrides to north–east Ireland.

In Iceland the thickness of Tertiary and Recent lavas is over 7500 m but it is unlikely that the Antrim Lavas ever came near that total, and an original thickness like that in Mull—some 1800 m—is more likely.

The earliest volcanic rocks, mainly basalt lavas with some beds of consolidated ashes (tuffs), were poured out over the land surface in a series of eruptions which formed a great plateau extending far beyond the limits to which its remnants are now confined. It seems probable that it covered at least most of north Derry, north Down, and Armagh and east Tyrone and built up a lava pile of great height. In spite of prolonged denudation since the eruption ceased, a thickness of 800 m is known in south Antrim.

The eruption of the lavas was followed by the upwelling of an acid magma in the Mourne and Carlingford areas. In the former, the magma was intruded below the sedimentary rocks and did not break surface, but in the Carlingford–Slieve Gullion areas, as in the great volcanoes of Skye, Rhum, Ardnamurchan, Mull and Arran in Scotland, the magma broke surface to give explosive volcanoes which were associated with complexes of intrusive rocks. Probably associated with these events was the eruption of acid lavas (rhyolites) at Tardree and the intrusion of a lopolith at Templepatrick, in mid-Antrim, which were in turn succeeded by further outpourings of basalt lavas, but the prolonged activity at the Slieve Gullion and Carlingford centres may have continued long after the lavas ceased to flow.

Late in the volcanic history came the intrusion of great masses of basic rock as sills or laccoliths, generally intruded into the underlying sedimentary rocks but occasionally into the lavas. Tensional cracks in the crust gave lines of weakness along which vertical sheets of basalt welled up to form dykes, often in swarms aligned north–west which converge on the major foci of the Mournes and Slieve Gullion.

Finally, the settlement of the crust caused large-scale warping and faulting throughout the area of the lava plateau, particularly round Lough Neagh where a deepening basin was simultaneously filled in Oligocene times with the series of sands, clays, and lignites known as the Lough Neagh Clays. Faulting in this area continued for a long time, for the Lough Neagh Clays themselves are faulted in Tyrone. Some of the old Hercynian faults were reactivated during this period—for example the Ballycastle–Dungiven Fault and the Annaghmore Fault in Tyrone.

The sub-aerial erosion of the land surface in Tertiary times has been the main factor in the development of the present-day land forms. The Cretaceous rocks which doubtless extended over most of Ireland were almost completely removed, except under the lava plateau in Antrim, and the rivers which developed on the Cretaceous cover are now superimposed on older rocks. The cover of Silurian rocks was stripped from the Mourne Mountains, and Tertiary planation surfaces have been recognized at about 250 m, 310 m, and higher levels up to over 600 m, in the Antrim hills, the Sperrins, Donegal, and the Mournes.

12. Tertiary: Extrusive Igneous Rocks

Lavas

With few exceptions all the lavas of the Antrim plateau are basaltic rocks, mainly olivine-basalts. They were erupted on the land surface and the tops of individual flows are commonly weathered to a red lateritic soil on which vegetation grew. Relics of this vegetation cover, preserved as lignite, contain spores of *Pinus,* a genus normally associated with temperate climates. The mode of eruption of the lavas is not altogether clear, though it is probable that they flowed both from fissures and from volcanic vents of classical type, the remains of which are common round the plateau area. Fissure eruptions are known to have occurred in historical times in Iceland, but in Antrim only one or two examples of a fissure feeding a lava flow are known. It is possible that in many cases the mass of molten lava would tend to flow away as a mobile pool when the eruption ceased and direct connection with the feeder would thus be broken. There is, however, no doubt that many of the volcanic vents also acted as sources of lava, particularly Slemish, where the traces of intermittent lava eruption can be seen in the plug, and it may be suggested that the whole lava plateau is possibly the relic of a much more extensive volcano of Hawaiian type.

Around the Antrim coast excellent cliff sections show the lavas piled one upon another in flat or gently dipping sheets. Individual flows vary greatly in thickness from thin slaggy beds only a metre or so thick to massive flows 30 m or even 45 m thick. In general the thick flows can be traced for greater distances, exceptionally for up to 10 km or more, while the thinnest flows are often seen to die out within a few metres, but some thin flows are very persistent while thick flows are often of quite limited extent.

Each flow consists typically of a basal slaggy layer, often with pipe-amygdales (Plate 9B), formed by steam bubbles rising from the underlying soil, overlain by a solid central portion which passes upwards into another slaggy layer, generally peppered with small gas cavities or vesicles. The upper part of this slaggy layer is frequently kaolinized by sub-aerial, and possibly ground-water, weathering to give a zone of purple-coloured rock, sometimes completely altered to lithomarge, and usually lateritized at the top to a red ferruginous bauxite. In addition to the typical tripartite flows there are a number of lavas of anomalous types. In several localities, particularly on the north Antrim coast as at Dunluce [C 906 415], banded flows up to 25 m thick can be traced for distances of several kilometres. These are apparently built up from a number of pahoehoe flow units, generally 2 m or so in thickness, each representing a period of lava spreading a few hours, days or weeks apart during the same eruption. The flow units are characterized by extremely abundant vesicles, the upper part of each unit being pale and decomposed and often topped by thin reddish bands representing rapid weathering of the glassy crust (Plate 8A).

Most of the lavas are olivine-basalts but tholeiitic basalts, virtually free of olivine, and fine-grained trachytes and mugearites occur locally. Acid rhyolite lavas are known from the Tardree area. From several localities—Cranny Water [D 255 178], Islandmagee [D 453 028], Lurigithan [D 220 228] and Skerrywhirry [D 285 006]—composite flows, consisting of layers of picrite-basalt and olivine-basalt, have been recorded.

Some of the flows which are exposed in the coastal cliffs appear to be ball-lavas, and it is possible that some beds which have been recorded as agglomerate may be actually lavas of this type. As described from Hawaii ball lavas consist of spheroidal lava balls in a vesicular matrix, formed by the accretion of solid lava around a core of clinker or a solidified bomb.

The solid compact part of the lava is invariably jointed, usually with random and irregular joints, sometimes curved, and fairly commonly with more or less regular vertical joints which give the lava a columnar appearance. In exceptional cases the vertical joints are so straight and so orientated at 120° between each set that the columns are vertical hexagonal prisms (Plate 10B). These are most common in the fine-grained tholeiitic basalts in north Antrim but rare in the coarser olivine-basalts, though they do occur, particularly where a hollow in the sub-basalt surface has produced a pond of lava which has cooled more slowly than usual. A good example of this is seen at Whitehead [J 474 913].

The texture of the rock forming the lavas varies from coarse-grained dolerite to very fine-grained basalt, mugearite and trachyte in which individual minerals cannot be seen with the unaided eye. In some of the finer-grained types, orientation of the minerals gives a recognizable flow structure, and even in the coarser olivine-basalt flow banding is fairly common, being seen in the parallel ribs etched out by the weather on some exposed surfaces. Coarse pegmatites are occasionally seen in the thicker flows, forming irregular sheets and veins, generally of no great thickness. These are late stage differentiates formed during the cooling of the molten rock and consist mainly of pyroxene and iron ore, feldspar, and abundant zeolites. They are well displayed in a cliff-top quarry west of Ballycastle [D 093 431].

Tuffs

Explosive activity at volcanic vents gave extensive beds of volcanic ash at several centres. Consolidated by heat and pressure these beds of tuff are a conspicuous feature of some of the northern cliffs. They are of maximum thickness in the vicinity of the vents from which they were ejected, reaching over 60 m at Carrickarade and 45 m near Kinbane. Tuffs ejected from these vents are found at horizons low in the lava succession and can be traced for about 7 km from Ballintoy to near Ballycastle. Less extensive deposits are seen at Ballycastle Harbour and near Ballygalley, both low in the lava succession, and at Skerriagh on Rathlin Island, low in the tholeiitic basalts. Thin tuffs, lateritized and almost indistinguishable from normal inter-lava weathered tops, occur in the lava succession on the foreshore west of Portrush and the pale interbasaltic bauxites of the Agnew's Hill area are apparently kaolinized rhyolitic tuffs.

The tuffs consist of comminuted fragments of basalt lava and of the sedimentary rocks through which the explosive vents were drilled. The most conspicuous constituent is usually Chalk, as at Kinbane; Lias shales are found in the Carrickarade tuffs and red laterite is generally plentiful. The fragments range in size from dust up to large blocks several metres in diameter. No true volcanic bombs have been recorded and much of the explosive activity presumably took the form of a gas release, unaccompanied by molten lava.

Interbasaltic Beds

The spasmodic eruption of the lavas and ashes was punctuated by quiet periods long enough to allow weathering of the tops of most of the individual lava flows and ash beds with the production of thin beds of laterite. There was also, during the volcanic epoch, at least one very prolonged period of quiescence during which the uppermost lavas were deeply weathered. The resultant thick bed of decomposed rock is recognizable over most of Antrim and in boreholes west of Lough Neagh. It is probable that during the period of its formation the lull was interrupted in two areas by eruption of lavas of different type. In north Antrim the group of tholeiitic basalts was poured out on to the weathered surface and was itself subjected to subaerial weathering to such an extent that in some areas the uppermost flow of this series gave rise to an upper interbasaltic bed of considerable thickness. In the Tardree area a series of acid lavas—rhyolites and glassy obsidian—formed a pile which may have been only partly covered by the subsequent basalt lavas, and which now crops out over an area of several square kilometres.

The weathering of the lavas took place in two stages. Firstly kaolinization of the rock, in which the oxidation of feldspar, pyroxene and olivine yielded silica and alumina which combined to form kaolinite. Most of the alkalis, magnesium and calcium were lost in solution and the resultant lithomarge consisted mainly of kaolinite and iron ore minerals. The secondary laterization process eliminated much of the silica and the alumina formed the trihydrate, gibbsite. Iron and titanium remained in the rock, the former largely oxidized to the ferric state, and concretionary and segregative processes tended to separate these oxides from the gibbsite with the formation of an iron-rich laterite at the top of the bed passing down into an aluminous laterite, or bauxite.

Where this process has been able to proceed to completion the iron concentrate is sometimes sufficiently rich to be mined as an ore. Old mines are numerous, especially in Glenravel. Weathering of basalt gave a ferruginous bauxite while rhyolite material yielded a siliceous bauxite, sometimes with quartz grains, as at Agnew's Hill [D 328 025].

There is little evidence in Antrim of the reworking of the laterites to give true sediments of Interbasaltic age, but beds of lignite and plant impressions are common at this horizon and lignite 6 m thick at Craigahullier [C 880 390] indicates a considerable accumulation of water-transported vegetation in a hollow in the weathered surface of the Lower Basalts. Thin beds of lignite have been worked as fuel at several localities throughout Co. Antrim, notably at Ballintoy. It may be noted that lignite always overlies grey bauxitic lithomarge, never red laterite.

Evidence from plant remains as to the age of the Interbasaltic Period has been somewhat contradictory and suggested ages range from the Eocene

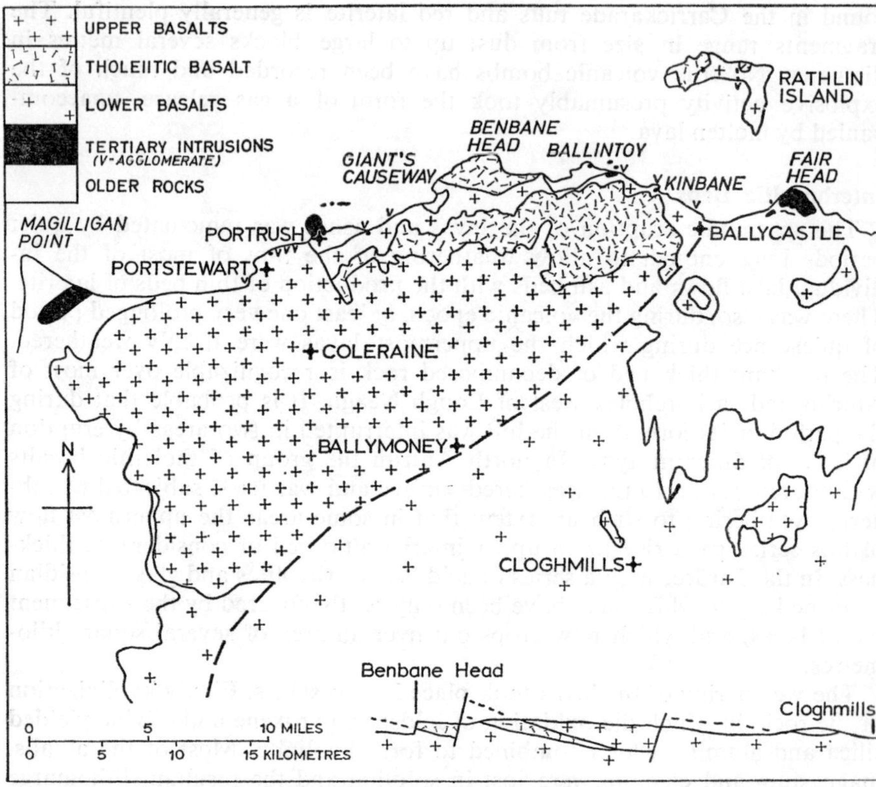

FIG. 24. *Distribution of basalt lavas and major Tertiary intrusions in north Antrim*

to the Pliocene. The most recent work on plant pollen suggests that they are Early or Middle Oligocene. Potassium-argon dating of the basalts indicates that they are Eocene or Palaoecene, about 60 million years old.

Volcanic History

In north Antrim the earliest volcanic activity was explosive, with vents at Carrickarade, Kinbane and Ballycastle ejecting volcanic ash which formed extensive tuff beds below and interbedded with the lowest lavas. The subsequent eruptions of olivine-basalt lavas built up a pile of Lower Basalt lavas which reached a thickness of about 90 to 120 m in this area. Away from the north coast the explosive activity was absent and the Lower Basalt succession reached a thickness of 180 m at Glenariff, about 300 m in south Antrim, 480 m at Langford Lodge, and over 240 m west of Lough Neagh. The lavas exposed along the margin of the Antrim plateau from Garron Point to Lisburn are all of this group which consisted mainly of olivine-basalts averaging about 6 m in thickness, though exceptional flows reach a thickness of over 30 m. As these are the flows which are preferred for quarrying, the average thickness of lavas exposed in working quarries is much greater than the 6 m quoted for the whole. In the Coagh area of Co.

Tyrone a fluviatile conglomerate of basalt blocks with some flint and laterite debris occurs near the base of the lava succession.

As well as olivine-basalts the succession includes occasional flows of trachytic and mugearitic types, seen at Glynn Hill and Slate Hill, near Larne, and at Ballinderry and Crew Hill, north–west of Lisburn.

In the area between Portrush and Ballycastle, on the north coast, the Lower Basalts are succeeded by a distinctive group of Tholeiitic Basalts, extruded during the Interbasaltic Period (Fig. 24). These rocks are hard, very fine-grained, break with a conchoidal fracture, rarely contain any megascopic grains of minerals and have little or no olivine. They occur in lava flows which are typically 20 m or more thick and which have a tendency to columnar jointing, which locally, in the lower parts of the flows, is very regular. Typical tholeiitic lavas have a three-tiered structure with a regular columnar base, a zone of slaggy or starchily columnar lava, and an upper zone of massive lava, in many cases showing rude columnar structure. They have few vesicles. Inter-lava laterites and lignites are common and a tuff bed occurs between two of the lower flows on Rathlin Island. Seven flows of Tholeiitic Basalts occur on Rathlin, nine are known at White Park Bay, and they die out westwards until cut out by a fault south of Portrush. Their greatest thickness is probably over 150 m.

The most spectacular development of the Tholeiitic Basalts is at the Giant's Causeway where a deep hollow, perhaps a river valley, in the Lower Basalts has been filled by the first flow of the tholeiitic series. The slow cooling of the great pool of molten rock, about 100 m deep, allowed the

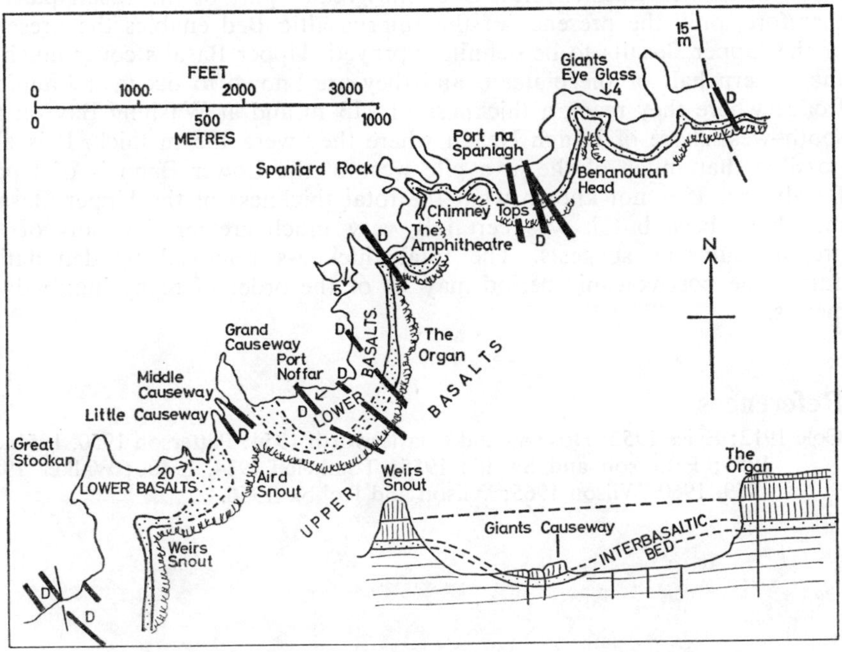

FIG. 25. *Sketch-map and section of the Giant's Causeway area*

formation of unusually perfect columns in the lower part of the mass, and it is this portion which has now been exposed by erosion. The 'Causeway' itself is the lowest part of the columnar zone and the purple lithomarge of the Interbasaltic Bed can be seen dipping beneath from both east and west (Fig. 25). It is dissected into three parts—the Grand, Middle and Little Causeways—by marine erosion along two dykes.

Probably contemporaneous with the north Antrim tholeiites were eruptions of acid rhyolite magma in the Tardree area and at a number of other localities near Ballymena and Templepatrick. The Tardree rhyolite is associated with porphyritic pitchstone or obsidian and rocks of these types now crop out over an area of 30 sq km. It is probable that they were normal lavas but agglomerate of pitchstone and rhyolite blocks fills a volcanic vent at Sandy Braes. In the Templepatrick area a bore penetrated a thickness of 430 m of rhyolite and it is possible that here there is a large intrusion of rhyolite in which the small chalk outcrops are xenolithic. Interbasaltic rocks derived from decomposed rhyolite are known at Agnew's Hill, near Larne, near Newtowncromelin and near Ballintoy. Some of this material may have been water-transported but it is probable that some of it was rhyolitic tuff.

The Tholeiitic Basalts and the upper leaf of the Interbasaltic Bed which overlies them, are succeeded in north Antrim by the Upper Basalts. South of Armoy the Tholeiitic Basalts probably die out—possibly by contemporaneous movement along the Tow Valley Fault—and the two Interbasaltic Beds come together, so that throughout the rest of the country the single Interbasaltic Bed is overlain by Upper Basalts which are indistinguishable from the Lower Basalt series. Over the greater part of the basalt plateau, therefore, only the presence of the Interbasaltic Bed enables the presence of the Upper Basalts to be definitely proved. Upper Basalts cover much of the eastern half of the plateau, and they are known to occur at Langford Lodge where they reach a thickness of 213 m and at Washing Bay on the south–west shore of Lough Neagh where they were 135 m thick. It is also possible that much of the lava pile west of the Lower Bann is of Upper Basalt age. It is not known what the total thickness of the Upper Basalts may have been but it was certainly very much greater than any of the present outcrops suggests. The total thickness removed by denudation during the post-volcanic period may be of the order of many hundreds of metres.

References

Cole 1912; Eyles 1952; Hospers and Charlesworth 1954; Patterson 1950, 1951a, c, 1955; Patterson and Swaine 1955; Tomkeieff 1934, 1940b; Walker 1951, 1959, 1960, Wilson 1965; Wilson and Robbie 1966.

13. Tertiary: Intrusive Rocks

Intrusive Complexes

Slieve Gullion

The south–west end of the Caledonian Newry Grandiorite was the site, in Tertiary times, of a revival of igneous activity which took the form of a ring complex along an arcuate fissure, coinciding closely with the margin of the earlier granite over much of its length, and a central mass of basic granophyric rocks forming a north-westerly trending range of hills which intersects the ring at its south–east side. The ring complex forms a circle of hills 11 km in diameter rising in places to over 300 m while the central mountain mass rises to over 510 m at Slieve Gullion itself and over 500 m at Clermont Carn in the south–east. The broad hollow between ring and central mass is floored by Newry Granodiorite (Fig. 26).

The ring complex includes a variety of rock-types. Relict masses of basalt and trachytic lavas downfaulted into vents suggest that the eruption of lavas was the first phase of the igneous activity. In the Forkhill area a series of volcanic vents extending over 10 km are filled with agglomerate consisting almost entirely of the local Newry Granodiorite and Silurian greywacke with some other pre-Tertiary rocks and some basalt and trachyte. A large mass of porphyritic felsite is intruded into the vent series and forms a belt several kilometres long and up to 1.5 m wide around the ring. A narrow outer ring-dyke of similar felsite forms a sharp ridge.

Round the north–western quadrant there is a ring of brecciated country rock, mainly of Newry Grandiorite and largely a crush breccia, though sometimes explosive breccias are found. Intruded inside the arc of breccias and extending round the rest of the ring, except where interrupted by the north-westerly ridge at Anglesey Mountain, is the main ring dyke of porphyritic granophyre, which in the west slightly overlaps the felsite ring dyke. This was the last intrusion of the ring complex and was itself formed by multiple injections.

The central mountain pile—the Central Complex—post-dates the ring dyke and consists of a layered mass of basic and acid crystalline rocks, in which thirteen layers have been distinguished. The whole was long regarded as of plutonic origin, the general opinion being that the Caledonian granodiorite was intruded by sill-like masses of dolerite and granophyre though the relative ages of these intrusions were uncertain.

An alternative explanation by Dr. D. L. Reynolds postulates that the layers are almost all extrusive in origin, consisting of a pile of rhyolitic and basaltic lavas (the latter sometimes very amygdaloidal and, rarely, in the form of pillow-lava), tuffs and agglomerates. These extrusive rocks have been subjected to metamorphism *in situ* during the subsequent history of the caldera which is assumed to have developed within the ring-dyke, and to have been transformed into granophyric and microgranitic rocks of varying composition, in the case of the acid lavas and tuffs, and into gabbro in the case of the basaltic lavas. Two of the basic sheets are, however, accepted as intrusive. The whole layered mass is penetrated by ramifying

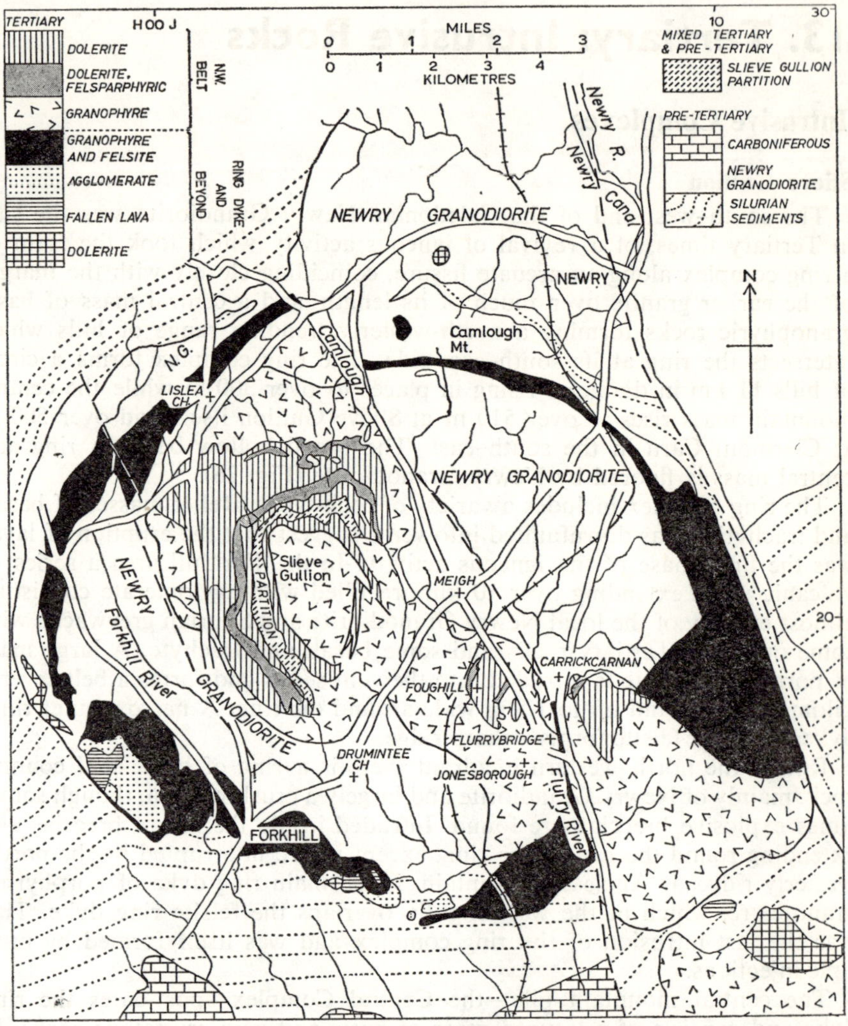

FIG. 26. *Sketch-map of the Slieve Gullion Complex*
(After Richey, Reynolds, Bailey and McCallian).

veinlets of fine-grained granophyre which are ascribed by Reynolds to the intrusion of incandescent acid tuff.

This hypothesis has not met with general acceptance. Bailey and McCallien explain the observed phenomena by assuming that the Caledonian granodiorite was invaded first by Tertiary granophyric magma which formed sheets separated by a layer of granodiorite (much recrystallized and contaminated by the intrusion). Subsequent invasion of the cool acid magma by sheets of basic magma caused the formation of basic 'pillows', simulating pillow lavas, and rheomorphic melting of the acid magma by the dolerite caused the formation of the transgressive granophyre veins.

A. *Sea-stacks and cliffs in Lower Basalt lava flows. West Lighthouse, Rathlin Island.*

PLATE 9

B. *Pipe amygdales in Lower Basalt lava. Black Head, Island Magee.*

A. *Regular columnar 'colonnade' and rubbly 'entablature'. Tholeiitic Basalt lava. Ballynastraid, White Park Bay.*

PLATE 10

B. *Polygonal jointing in Tholeiitic Basalt lava. Giant's Causeway.*

A. *Volcanic neck rising from a plateau of basalt lava. Slemish.*

PLATE 11

B. *Volcanic explosion about* 10 *m wide cutting through basalt lavas. Devil's Port, Portrush.*

Inclined volcanic neck. Tievebulliagh. The neck is the dark mass lying on the Lower Basalt scarp. Neolithic excavations for porcellanite, used in the local axe factory, can be seen in the scree below the neck. The small excavations round the flank of the hill are old Chalk workings.

PLATE 12

The difficulty found by the early observers in deciding the relative ages of the intrusions was due to the fact that both acid and basic rocks are seen to be invaded by, and chilled against, each other with the production of hybrid contact rocks. While Reynolds ascribes the hybrids to fusion of the wall rocks by the incandescent tuff streams, the other interpreters suggest the chilling of basic magma against still fluid or remelted acid magma.

Carlingford

Although lying entirely in the Republic of Ireland, this brief account of the plutonic complex of Carlingford is included as it is one of a trio of Tertiary intrusive centres, separate but closely associated in location and time. Like the Mournes, this complex is emplaced in Silurian shales and greywackes, but the southern and eastern part is intruded into the Carboniferous outlier which overlies the Silurian strata from Greencastle to Dundalk (Fig. 27).

As at Slieve Gullion, with which it was doubtless contemporaneous, the earliest activity was eruptive, with the extrusion of tholeiite and alkaline basalt lavas now exposed on the south side of the complex near Rampark,

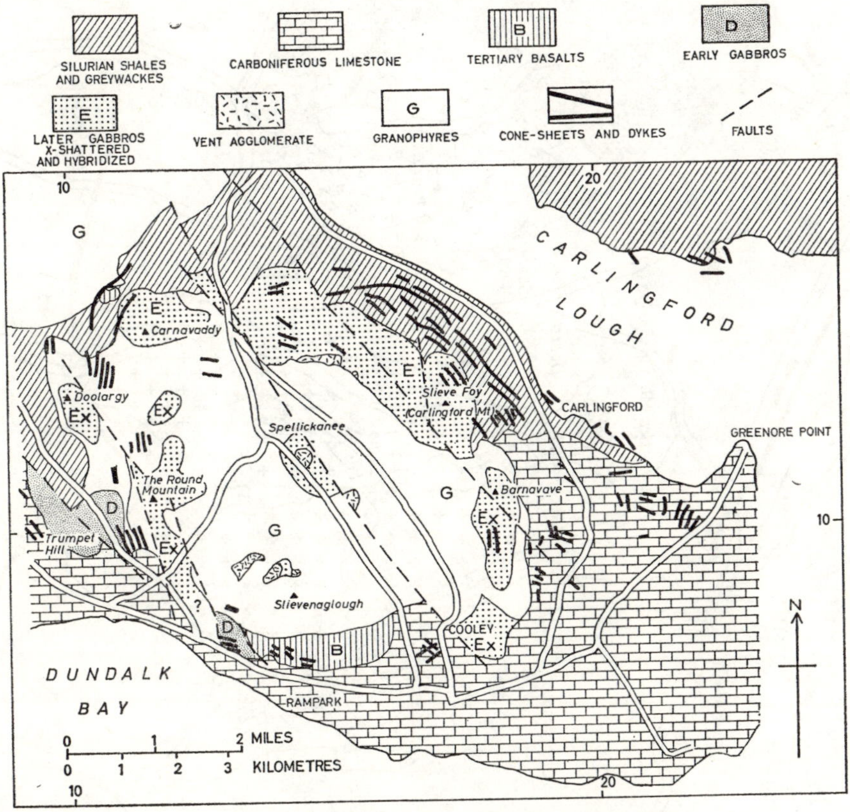

Fig. 27. *Sketch-map of the Carlingford Complex* (After Le Bas).

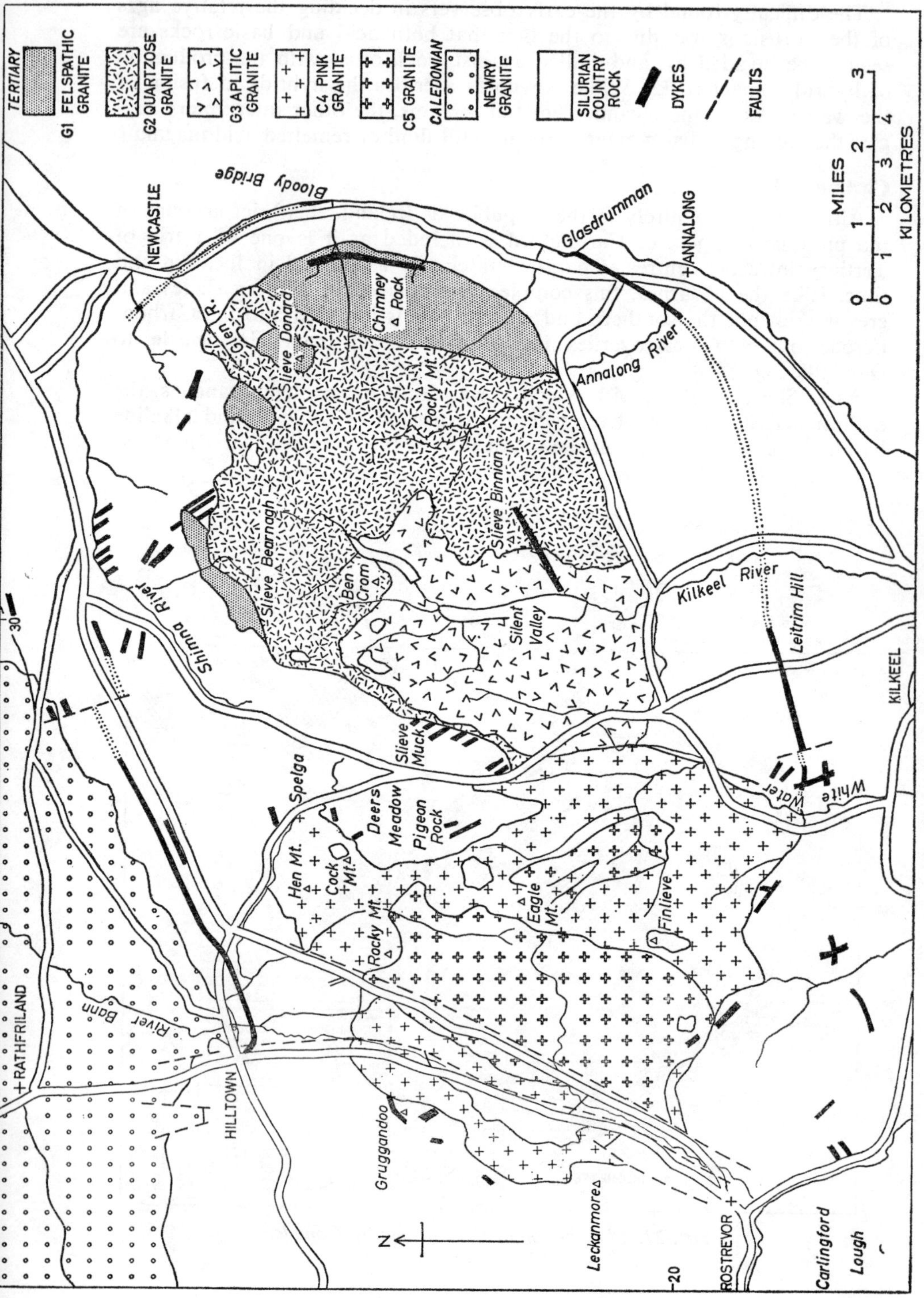

FIG. 28. *Sketch-map of the Mourne Mountains*
(After Richey and Emeleus)

where they are about 275 m thick. This was followed by the intrusion of a number of elongated plugs aligned north–west—the Early Gabbros—and a lopolith of olivine-gabbro, generally hybridized by the subsequent granophyres except where it overlies the Silurian rocks and has not been affected by the later intrusions. The lopolith—the Later Gabbros—is now seen as a series of relict outliers on the hills around the periphery of the complex and on two hills in the middle of the area. At their best preserved development on Slieve Foye, the Later Gabbros are over 365 m thick and can be divided into four layers each marked by olivine segregation at the base and variation in feldspar composition from base to top. The gabbros contain xenoliths of Silurian rocks, altered to pyroxene-hornfels, and Carboniferous limestone, recrystallized with development of abundant calc-silicate minerals.

The Gabbros were followed by the intrusion of granophyres at two periods separated by a volcanic episode. The earlier granophyre is known only from fragments in agglomerate of the middle period which is seen in vents well exposed on Slievenaglogh. The main granophyre body is regarded as intruded in the later period and occupies an area of about 30 sq km in the middle of the complex. Where it is seen to intrude the Carboniferous Limestone, the latter is recrystallized with the development of melilite, wollastonite, diopside, forsterite, garnet, cuspidine, monticellite, phlogophite, and spinel.

Centred on the complex is a set of radial basic dykes and abundant cone sheets. The last intrusive phase was the development of a north–west swarm of basaltic dykes.

Mourne Granites

The granite mass of the Mourne Mountains (Fig. 28) now exposed over an area of about 150 sq km, intrudes Silurian sediments which are disturbed and thermally hornfelsed for short distances from the contact. Remnants of a roof of the sediments are seen at several places and it is clear that the granite was intruded at a high level in the crust but did not reach the surface. The granite truncates some trachytic and dolerite dykes of earlier Tertiary age—the result of crustal weakness over the magma-reservoir before the emplacement of the main granite intrusions. Geophysical evidence suggest that the granites are relatively thin and are underlain by a mass of basic rock. Basic xenoliths are, in fact, found in some of the dykes.

The granite has been emplaced in successive injections at two centres. The eastern centre has three distinct granites—a feldspathic hornblende-granite, G_1, a quartz-rich granite with abundant dark quartz, G_2, and an aplitic fine-grained variety, G_3. The western centre has a pink outer granite, G_4, enclosing a later fine-grained microgranite or granophyre, G_5.

In the eastern centre the granites are sufficiently eroded to show bounding walls as well as roofs and the mode of emplacement is clear. The three intrusions are arranged within one another with marked eccentricity towards the south–west and the intrusions were at sufficiently long intervals to permit each mass to cool and solidify and to chill the contact margins of the succeeding intrusions. There is no evidence of extensive uplift of the country rocks, and the abrupt turn of wall into roof in the case of each

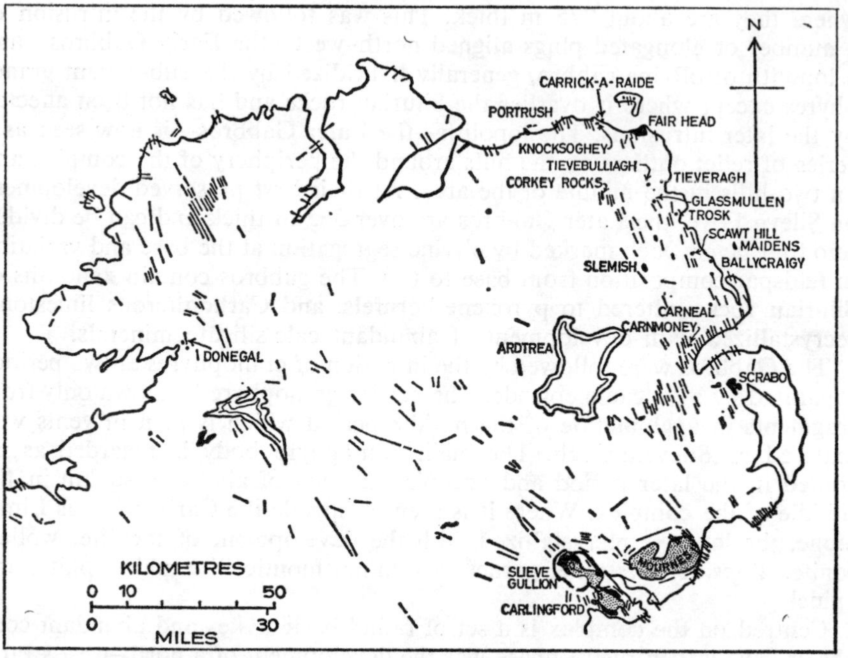

FIG. 29. *Location map of Major Tertiary intrusions and dykes in Ulster.*

intrusion suggests that the successive emplacements were due to the subsidence of a block of the country rock bounded by a ring fracture and separated from the roof by a cross fracture, with the influx of magma from below into the ever-widening fissure and the roof space.

The precise age-relationship of the granites of the western centre to those of the east is uncertain though it is probable that they are later. The mechanism of the intrusion of the western granites is similar to the eastern, and at the present level of denudation no wall boundary is exposed, the domed surfaces of G_4 and G_5 representing flat-lying cross fractures.

Acid and composite cone-sheets are seen at several points round the granites, being well exposed at Glasdrumman Port [J 379 223], Hilltown [J 217 287], Gruggandoo [J 205 260], and Formal [J 234 166]. Numerous aplitic sheets throughout the granites represent the intrusion of late-crystallizing fractions of the magma into cracks in the consolidated granite. The nature of the aplite is closely controlled by the variety of granite into which it was intruded, but sometimes older granites are invaded by later aplites near the contacts with the newer granites.

Basic Intrusions

Plugs

Vertical and inclined magma-filled pipes, varying from circular to extremely elongate in plan and ranging from 50 m to 1 km in maximum diameter, are known at over thirty localities within the limits of the basalt plateau and its immediate periphery (Fig. 29). Elsewhere in Ireland they are virtually unknown and only isolated examples near the town of Donegal and at Bunowen in Connemara are recorded. In Antrim they usually stand clear of the surrounding lavas as steep raised prominences and there is a tendency for the elongation of plugs to be N.N.W.–S.S.E., parallel to the general direction of the regional dyke swarm. Some of the plugs lie on fairly well defined lines which may be rift zones. The plugs are generally composed of olivine-dolerite, locally with an olivine-rich phase near the margins, due probably to differentiation in the ascending column, which may also account for the composite flows of lava. Such picritic margins are found in Tieveragh [D 231 284], Tievebulliagh [D 193 268] and Corkey Rocks [D 092 224], while Trosk [D 272 203] is entirely picrite-dolerite. Carnmoney [J 333 825] and Glassmullen [D 215 224] are olivine-free dolerite.

Of exceptional interest are the plugs at Slemish [D 222 055] and Tievebulliagh. Slemish, the largest Irish volcanic vent, with an oval plan a kilometre across, was formed by multiple intrusion (Fig. 30). The conduit was filled at lower levels by at least two intrusions D_1 and D_2, while the final intrusion, D_3, broke through the lower solid plug and fills the conduit in the upper part of the hill. D_1 and D_3 contain cognate platy xenoliths which represent the foundered fragments of surface crusts from lava lakes formed in the crater and distributed by convection currents in the lava. The plug

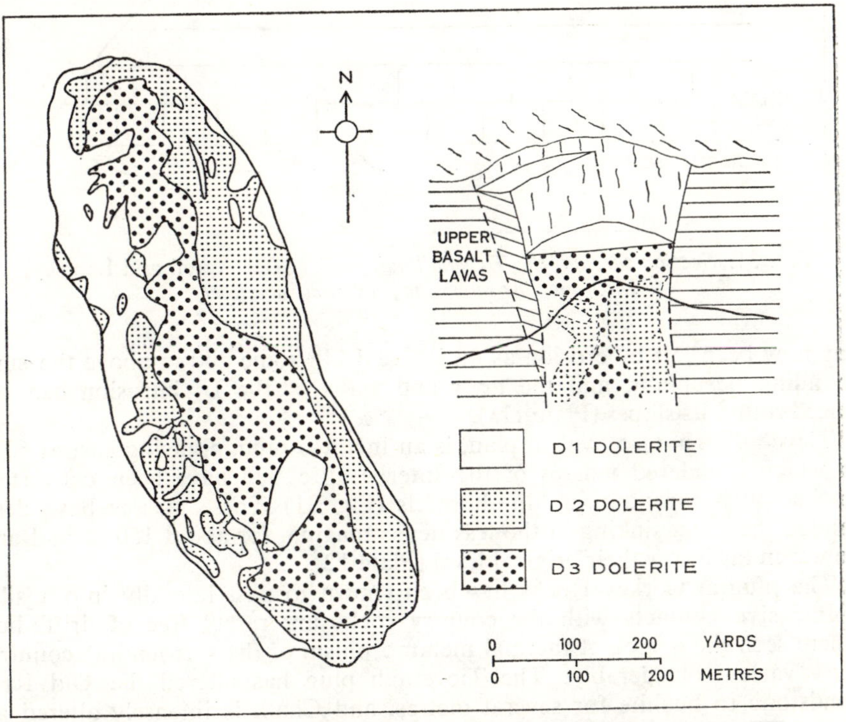

FIG. 30. *Sketch-map of the Slemish volcano and reconstruction of the volcano during the lava lake phase of the third intrusion.* (After Preston).
The heavy line shows the present erosion level.

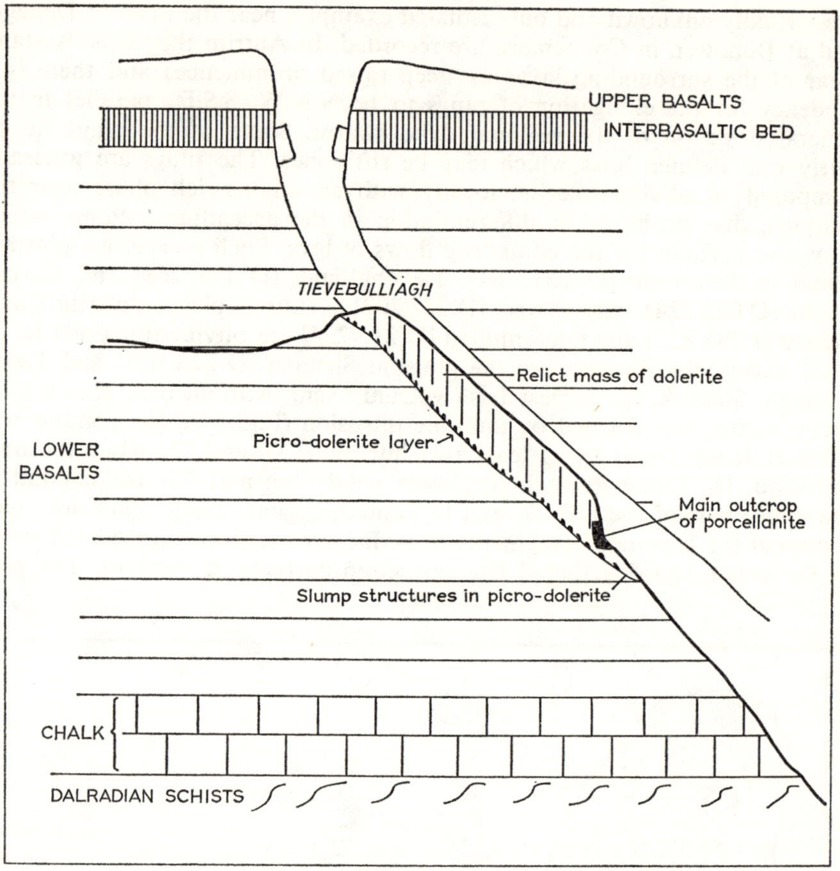

FIG. 31. *Section through the Tievebulliagh plug.* (After Agrell and Langley) *The heavy line shows the present erosion level.*

has now been left by erosion as an isolated hill some 200 m above the surrounding peneplain, and the floor and walls of the D_3 intrusion can be traced round its slopes (Plate 11A).

Tievebulliagh, also oval in plan, is an inclined plug, plunging east at 60°, in which foundered masses of the Interbasaltic Bed have been converted into a tough corundum-mullite-hornfels (Fig. 31). These masses have descended the plug, sinking in the less dense magma, for about 150 m and are now seen far below their original level (Plate 12).

The plug at Corkey Rocks has been shown to pass laterally into a sill.

Intrusive contacts with the country rocks are rarely free of drift, but where seen the degree of thermal metamorphism of the surrounding country rock varies considerably. The Tieveragh plug has altered the Old Red Sandstone to buchite for several metres, and Chalk is intensely altered at Scawt Hill [D 337 091], Ballycraigy [D 387 045] and Carneal [J 389 959] with the production of a wide range of metamorphic minerals such as scawtite, larnite, spurrite, wollastonite, etc. Tievebulliagh has profoundly altered the adjacent basalt lavas, and as Brockley [D 118 520] an interbasaltic laterite has been altered to a dense corundum-mullite-hematite-hornfels. This rock and the similar hornfels from Tievebulliagh are of archaeological importance having been used by Neolithic man to make widely distributed axes.

Two of the plugs—Ballygalley Head [D 383 078] and Carnmoney—are associated with agglomerate wedges, apparently relics of tuff-filled vents which were subsequently refilled with magma.

Agglomerate Vents

There were two major centres of explosive activity in north Antrim—the vents of Carrickarade [D 061 449] and Kinbane [D 088 438]. The former, well exposed in a classic coastal section, lies towards the east edge of a large ash cone which reaches a thickness at Knocksoghey of 60 m. The vent, 275 m in diameter, is probably only peripheral to a larger vent, below sea level in the area north of Larry Bane Head. The agglomerate in the vent and the tuff in the ash cone are largely chalk but with much basaltic material and some Lias debris. Vents in the Carrickarade area were active at the beginning of the volcanic period—a tuff bed overlies the Chalk, beneath the earliest basalt lava, in this area—but the date of the last eruptions at Carrickarade itself was later. The explosive activity was followed by irregular dolerite intrusions which form much of Carrickarade Island.

The Kinbane vent (Fig. 32) is probably also offshore with only the marginal features known. It was active at the beginning of the volcanic period, forming a tuff bed at the base of the lavas, but was also reactivated halfway through the Lower Basalt period, when a second series of tuffs was laid down. Like Carrickarade it has associated dolerite intrusions and considerable disturbance of the lavas and chalk is evident in the Kinbane district.

Between Ballycastle and Portstewart cliff and shore sections show agglomerate-like material in over thirty localities. Some of these are probably collapse breccias, formed by collapse of lavas into solution cavities in the underlying chalk, but over twenty are clearly of volcanic origin. It has been suggested that they may be due to phreatic activity—hot lava overriding water logged chalk giving steam explosions—but many of them are of true explosive volcanic type and have been compared with the Schwabian tuff pipes. A number are associated with beds of agglomerate. The small vents average about 20 m in diameter and exceptionally good examples are seen in the cliffs at Devil's Port [C 845 398] and Gortnabane [C 908 417] (Plate 11B).

An agglomerate-filled vent, with blocks of perlitic obsidian, is seen at Sandy Braes [J 205 959] in an area of rhyolite lavas. Other vents are known at Tullyloob [J 120 616], Rockville quarry [J 229 671], Scrabo [J 478 724], and Ballymena [D 102 021].

Sills

Large sills occur at a number of points around the lava plateau, generally intruded into pre-Tertiary sediments. Only two are known to be intruded

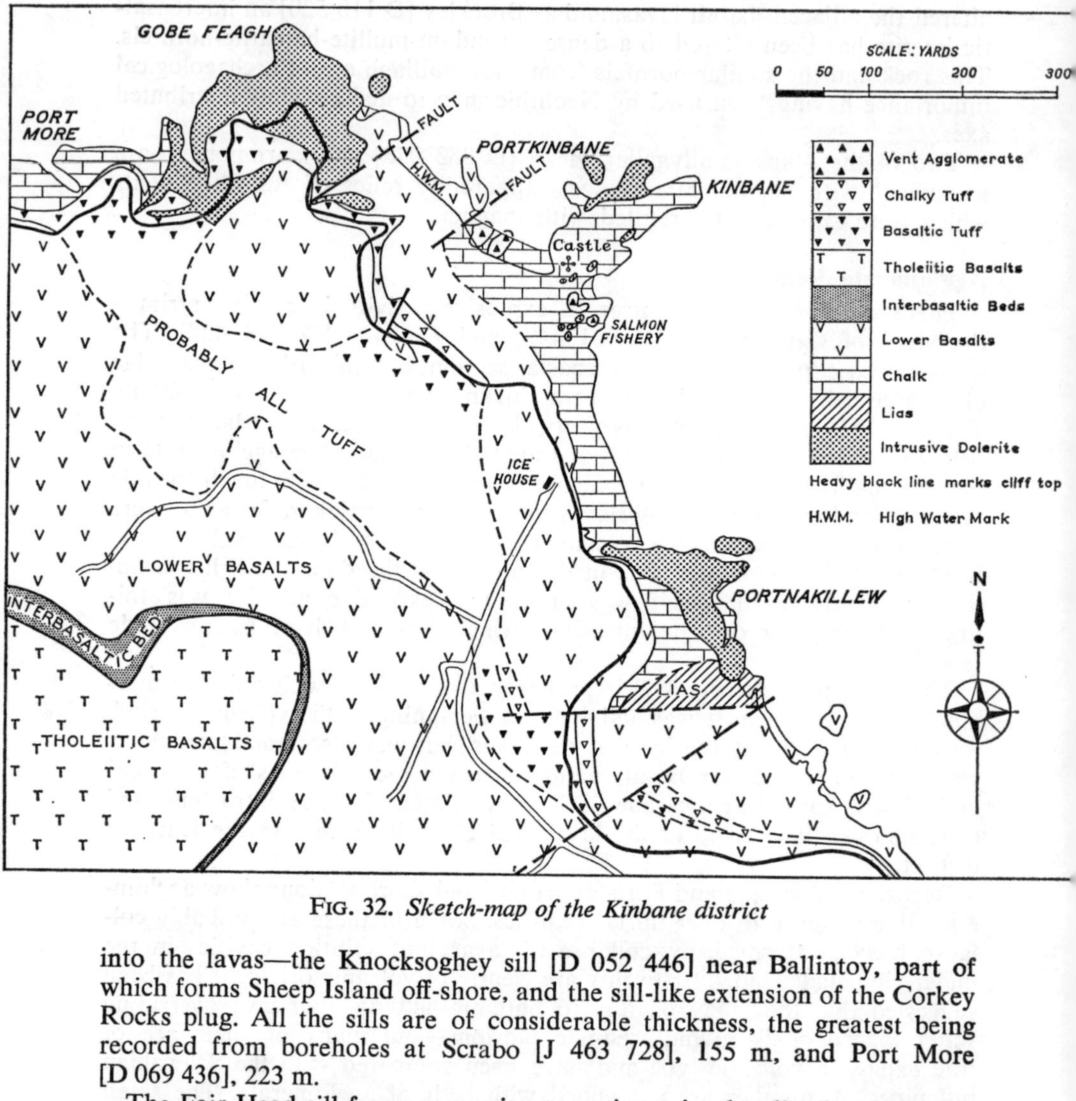

FIG. 32. *Sketch-map of the Kinbane district*

into the lavas—the Knocksoghey sill [D 052 446] near Ballintoy, part of which forms Sheep Island off-shore, and the sill-like extension of the Corkey Rocks plug. All the sills are of considerable thickness, the greatest being recorded from boreholes at Scrabo [J 463 728], 155 m, and Port More [D 069 436], 223 m.

The Fair Head sill forms a capping to an imposing headland and reaches a thickness of 82 m (Plate 13). It can be seen to thin to the south-east, transgressing upwards through Carboniferous, Triassic and Cretaceous sediments, and to the west, where it interdigitates with Carboniferous sediments [D 168 429]. Its upper part is extensively penetrated by dolerite-pegmatite sheets and veins, and veins of magnetite-rich basalt, and is cut by a few thin basalt dykes. In one area it has large xenolithic rafts of Carboniferous sandstone [D 187 430]. A complex of minor but extensive sills extend to the south and west at higher stratigraphical levels and a thick sub-parallel sill crops out at a lower level among the talus on the headland.

The Portrush sill, over 30 m thick with base not seen, forms the promontory of Ramore Head and an arcuate chain of islands known as the Skerries.

It is notable for the extensive occurrence of olivine-rich segregations which form distinctive dark clots in the rock, and for the occurrence of a remarkable series of late-stage syntectic veins of pyroxene-plagioclase, and veins of modified hornfels. Hornfels, formed by the intense induration of the Lias shales, can be seen to overlie the sill and contains recognizable fossils, particularly ammonites. This 'Portrush Rock' was the subject of controversy in the late eighteenth century between 'Neptunists' and 'Vulcanists'.

Little is known about the sill which forms the Maidens, off Larne, except that it is, like the others, an olivine-dolerite. The Ardtrea sill in Tyrone is 3 km long, and two sills, each over 1·5 km long, are intruded into the Triassic and Permian rocks at Kingscourt, Co. Cavan.

Far to the west an extensive sill crops out in the Carboniferous rocks between Garrison and Derrygonnelly. This intrusion underlies an area of 15 sq km and is apparently of Tertiary age. Even more remote from the main centres of Tertiary activity are large sill-like bodies at Killala Bay, over 5 km long, and in the Castlebar and Foxford areas where one intrusion in the Carboniferous rocks is over 8 km long, and others exceed 1·5 km in length of outcrop.

Tertiary Dykes

Most of the dykes of Tertiary age are relatively narrow flash injections which in the last stages of the igneous process exhausted the magma supply, but some are earlier and fed large volumes of magma to high-level intrusions or lava flows. These feeder dykes are usually wider (up to 30 m) and show signs of multiple intrusion with internal contacts marking successive surges of magma along the fissure. They also have often distinct metamorphic aureoles in the country rocks.

In eastern Ulster concentrations of basic dykes are seen, generally in coastal sections, round the northern and southern ends of the Antrim Plateau, in the Ardglass district and round the Mourne and Carlingford mountains. The latter three groups have been described as 'Dyke Swarms' and in the case of the Mournes and Carlingford areas they are clearly associated with the plutonic centres. The dykes of north Down and Antrim have likewise been grouped as Hillsborough and Tardree dyke-swarms but it seems at least possible that this classification is based on chance frequency of exposure and has little significance. Over 300 dykes are known in the Belfast area, indicating a crustal stretch of up to 4 per cent.

In western Ulster the distribution of known dykes is more random and they are much less abundant, though often they are much larger. The most remarkable concentration is in the central part of the Donegal Granite where an intense swarm of north–north–west trending dykes occupies a joint system in the Caledonian granite, and a similar concentration is seen with the Barnesmore Granite as focus. Apart from these areas dykes are relatively scarce in the metamorphic rocks of Donegal, Tyrone and Derry but there is a wide belt stretching across Fermanagh, south Tyrone and Monaghan where large feeder dykes, generally with a north–west trend are more abundant, and this group may be associated with a buried igneous mass which is suggested, on geographical evidence, to underlie south–west Tyrone.

Such dykes are well exposed in the quarries near Dora Villa [H 200 558] and on Magho Mountain near the scenic Road [H 080 560] where the dolerite is full of quartz xenoliths showing evidence of fritting and the growth of tridymite.

The dykes of Antrim and north Down are almost exclusively of olivine-dolerite or olivine-basalt with only rare olivine-free tholeiitic exceptions. They trend as a rule north–north–west though there are occasional anomalous directions including some east–west dykes which have been regarded as the earliest intrusions.

The Hillsborough–Ardglass–Killough swarm is mainly of olivine-basalt dykes but about one quarter are of trachybasalt. All contain abundant carbonate and analcime. The predominant trend is north–west but there are a few east–north–east examples.

The 130 dykes known on the 16 km of coast round the Mourne Mountains differ from the other swarms in their dominantly intermediate composition —only 19 per cent are of olivine-basalt, the remainder being olivine-free basalts and andesites with some acid porphyries and felsites. Composite and multiple-intrusion dykes and sheets are common. This swarm was assumed to be mainly pre-granite in age as no comparable concentration was recognized in the area underlain by granite and some of the dykes clearly stopped at the margin. A tunnel in the granite has, however, revealed thirty dykes in 4 km, of which two are basic, three composite, fourteen basic porphyrite, and eleven acid porphyrite. As the rock types in the groups described from without and within the granite area are comparable it seems probable that most of the Mourne dykes are later in age than the intrusion of the granite.

A concentration of basic dykes runs N.W.–S.E. through Slieve Gullion and Carlingford, and includes a few composite intrusions which are aligned with the large feeder dykes of Fermanagh and south Tyrone.

References

Agrell and Langley 1958; Bailey 1959; Bailey and McCallien 1956; Brown 1956; Charlesworth 1937; Charlesworth and Hartley 1935; Cole 1912; Dawson 1951; Elwell 1958; Emeleus 1955, 1962; Harris 1937; Le Bas 1960; Nockolds 1935, 1938; Old 1970; Patterson 1946, 1951; Patterson and Swaine 1957; Preston 1962, 1963, 1965, 1967; Reynolds 1931, 1934, 1943a, b, 1951, 1961; Richey 1927, 1932, 1935; Robbie 1955a; Sabine 1968; Tilley 1929; Tilley and Harwood 1931; Tomkeieff 1935, 1940; Tomkeieff and Patterson 1947, 1948, 1953; Tomkeieff and Marshall 1935, 1940; Walker 1959; Wilson 1964; Wilson and Robbie 1966.

14. Tertiary: Lough Neagh Clays

The end of the volcanic period was followed in north–east Ireland by a period of earth movements. Widespread warping of the crust gave a general synclinal form to the lava plateau, and was accompanied by faulting on a large scale. Much of the latter activity was centred on Lough Neagh and a combination of synclinal folding and block faulting on earlier structural lines in this area produced a tectonic basin. The denudation of the surrounding higher ground poured very large quantities of sediments into this hollow and probably within a relatively short period the basin was filled with the deposits now known as the Lough Neagh Clays. The end of the tectonic movements came before the end of the sedimentation, and in some areas the Lough Neagh Clays overstep on to the older rocks beyond the faults.

The present extent of these beds is some 500 sq km of which 300 sq km underlie Lough Neagh. The outcrop forms an irregular arc round the southern end of the Lough with greatest extent beyond its south-west corner. Because of their soft and unconsolidated nature the Lough Neagh Clays are low-lying and are generally concealed by surface deposits of glacial drift and peat and exposures are few. The beds have, however, been penetrated by several deep bores in the area east of Dungannon and shallower bores in the Crumlin area, and a good deal is known about them.

The greatest known thickness of the formation is at Washing Bay where 349 m were recorded and thicknesses of over 150 m are known from several other localities. While in general the beds consist of pale coloured sideritic clays with beds of sand and lignite, there are wide variations from place to place in the lithological succession and there is little doubt of their deposition by rivers carrying detritus from different sources.

The lowest beds are in places a conglomerate of basalt, flint, chert, Old Red Sandstone volcanic rocks, Triassic sandstone, and quartz grains, as near Coalisland, but generally the pebbles are scattered through some thickness of clayey beds. The remainder of the succession is dark clays, commonly lignitic and with local beds of lignite, tending to pass upwards into paler sandy clays with poorly preserved plant debris. The lignite was formed from timber washed into the lake with the sediments. These pale clays have been worked at various points on the western side of the Lough for ceramic manufacture and are still worked at Ballynakelly, near Coalisland, which is the only location where they are well exposed. The lignites which generally occur at depth in the deeper part of the basin approach the surface in the northern part of the outcrop east of Stewartstown where they were once worked, and more importantly, on the east shore of the Lough near Crumlin where they were fairly extensively worked at one period and are still visible in one exposure [J 133 740]. There is clear evidence that in the more massive lignite beds silicification of the timber took place. The silicified wood was recovered from the lignite pits and some of it is still extant. One large piece in the Ulster Museum shows the transition from normal lignite to complete replacement by silica.

The Lough Neagh Clays are very poor in fossil remains, except for plant debris. *Unio* and *Viviparus* shells were found in the Washing Bay Borehole but these are of no value for age determination. A few fossil plants have been named—*Sequoia couttsiae, Dewalquea hibernica, D. fraxinifolia,* and *Sphagnum*. Recent work on pollen from the lignite clays indicates that the deposit is probably of early or mid-Oligocene age.

Earlier suggestions that the material forming the clays came entirely from the weathering of the Antrim Basalts and rhyolites is now discounted and it seems probable that much of it was derived primarily from the denudation of the Carboniferous sandstones and fireclays to the west, with some contributions from all the other rocks in the surrounding catchment.

References

Barton 1757; Fowler and Robbie 1961; Scouler 1837; Watts 1962, 1963; Wright 1924b.

15. Quaternary

Like the latter part of the Tertiary era the beginning of the Quaternary was a period of denudation in the northern part of Ireland and no Pleistocene deposits of this age are known. Relics of a pre-glacial shoreline have, however, been recognized in Co. Down as a rock-cut bench at a height of about 4 m above present mean sea level. On the south coast of Ireland at Courtmacsherry and in the south of Great Britain this platform has relict beach deposits overlain by glacial or peri-glacial drift.

The Quaternary Ice Age affected the whole northern hemisphere and from about two million years ago climatic fluctuation has caused an alternation of glacial periods, during which ice-sheets covered wide areas, and interglacials during which temperatures were as high, or higher than, the present. Early this century four glaciations were recognized to have affected Central Europe but more recent evidence has suggested at least six cold periods. The main glaciations, Biber, Donau, Günz, Mindel, Riss and Würm—the last three known as the Elster, Saale and Weichsel in northern Europe—are recognized on the evidence of interglacial deposits with fossil plants and animals, and each of these stages can be subdivided by interstadials which marked shorter climatic fluctuations. There is no evidence that the varying conditions that led to the repeated climatic changes of the past have now ended and it is probable that at present we are living in a temperate interglacial period. Estimates of interglacial time are extremely unreliable but there have probably been at least 60 000 years in each episode. As the time since the retreat of the last major ice-sheet from Ulster can be estimated as about 18 000 years we may now be in the early part of an interglacial period.

The rapid development of pollen studies in the last decade has improved the reliability of wider correlation of glacial and interglacial organic remains, but perhaps the more important advances in technique have been absolute dating by means of radio-active isotopes—radiocarbon and potassium-argon—and palaeotemperature determinations. The latter, from ocean basin sediments, may give a general scheme of climatic change during the Quaternary which could become the basis for a world-wide correlation.

In Central Europe, and in particular in the Alps, few organic interglacial deposits occur which can be dated with certainty relative to the main Alpine glacial stages. As a result correlation over wide areas of glacial and interglacial stages is by no means clear. For this reason, local Quaternary successions have been established recently in Britain. Abandonment of the rigid and sometimes confusing Alpine nomenclature has, undoubtedly, resulted in advances in knowledge of local British Quaternary stratigraphy. Similarly sequences of deposits of various ages have been established in Ireland but here also organic deposits within the glacial successions are rare and there has been little local opportunity for the application of modern techniques of palynology and geochronology.

The local terminology established for use in Ireland and its correlation with Britain is shown in Table 4. The more important synonyms used in past literature are given and for the last glacial period a curve of climatic change as determined in north–west Europe is added to indicate the typical oscillations of climate.

In Ulster the general stratigraphy, or order of deposits, is now established but there is little evidence before about 10 000 years B.P. of the precise age of the sediments. As a result they can equally well be regarded as referable to the last (Midlandian or Ivernian) glacial period or as representing deposits of both Midlandian and Munsterian periods. The two views are represented in Table 3. Eventually improved reliability of correlation with Britain and the Continent may help, but the problem seems unlikely to be entirely resolved until at least one unequivocal interglacial deposit of the Ipswichian period is found. At present no material of this age is known from Ireland. The Ardcavan deposit, once widely quoted, has now proved to be of later age. Interglacial deposits are known in the southern part of Ireland but pollen evidence shows they are all of Gortian (Hoxnian) age, the type site being at Gort, Co. Galway.

The glaciation of the country effectively removed almost all previous unconsolidated deposits and left in its train a variety of drift material which accumulated during the movement of the ice-sheets and, in particular, during their decay when they were melting to give large quantities of water. The most widespread of these drifts is in the boulder clay—till, or ground-moraine—which formed below the ice, sometimes as an irregular sheet, sometimes as elongated mounds or drumlins (Fig. 33, Plate 14A). Mounded into stream-lined forms which offered the least resistance to the moving ice-sheet, the drumlins now reflect the directions in which the ice moved. During the melting or ablation of the ice-sheet interbedded silts, clays and sands were formed in, on and under the ice. These deposits include the ridge-like eskers, or osar, which were probably formed in subglacial stream tunnels during the waning stage of glaciation (Plate 14B).

On the periphery of the ice, mixed up with and beyond the end moraine, meltwaters deposited outwash sands and gravels including kames and kame terraces—undulating sand and gravel terrain with kettle-holes left by the melting of buried masses of ice. Near the margin of the ice the till may have become saturated with water and flowed as a slurry. Such flow tills can extend some distance into the well stratified outwash sands and gravels. A plain formed by outwash material, which may include flow tills, is distinctly seen around Limavady, Co. Londonderry, the town itself being built on outwash material. Varved clays were laid down in standing water in ice-marginal lakes and are one of the important proglacial deposits. Examples occur in the lower Bann Valley around Agivey [C 891 219], east of Gortin in the Owenkillew Valley and at Legnagappoge [H 538 967] in the upper part of the Burndennet Valley.

From the distribution and content of all these deposits it is possible to deduce the sequence of events during the ice movements and last phases of glaciation, but much of the evidence is equivocal and conclusions are often contentious.

Though the glacial period lasted for about two million years only the last events can be discerned. In Ulster four stages are well authenticated: (1) an ice-sheet flowing east or north from a centre probably located in Donegal (known only from local deposits in the Lough Foyle basin); (2) an ice-sheet from Scotland streaming south or west across Ulster; (3) an Irish ice-sheet moving east and north; (4) a readvance of Scottish Ice which moved into north Antrim and north-east Londonderry and down the North Channel.

The earliest ice-sheet (1) probably flowed east or north from a centre in Donegal. A centre to the west of the country where the westerly maritime air first meets high ground would be a natural area of precipitation and snow accumulation—in fact a similar situation to that found in the development of the Scandinavian ice-sheet. At the same time local ice centres may have formed in the Sperrin Mountains and the Mourne Mountains may have had a local ice centre, though the various early till-facies in this area are at present difficult to correlate with certainty in the remainder of Ulster. The earliest ice movement is proved by a distinctive basal till at Lackagh, Co. Londonderry [C 612 191] where the three main tills, resulting from the ice-movements (1), (2) and (3) described above, are exposed in order of deposition. This lowest till contains erratics dominantly of Dalradian rock-types and is correlated with a similar till described from a temporary excavation on Blanket Nook Levels, Co. Donegal [C 310 190].

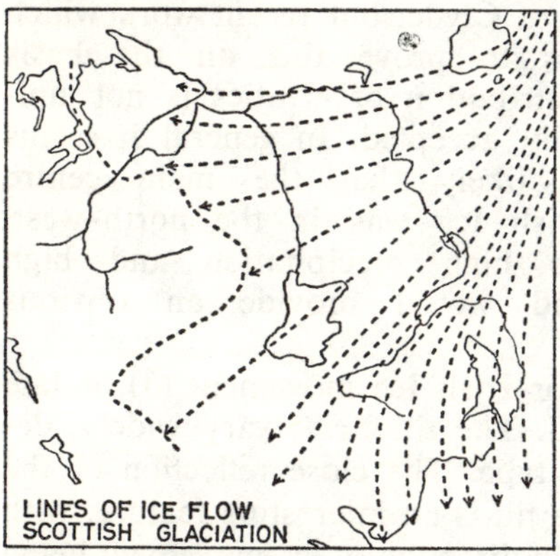

The early Scottish Ice (2) fanned out from the Grampians, moving south down the North Channel and across Antrim and Down, and south-west across the Lough Foyle area where it deposited a typically calcareous till with marine shells. This is still preserved in the low ground north of Loughermore Mountain and the type section of this till formation has been taken at Bovevagh [C 678 138]. The till was first correctly described in the Geological Survey Memoir on the Londonderry area and the name 'Londonderry Lower Till' is most apt. The extent of this Scottish Ice is shown by the occurrence of unique erratic rocks, particularly the riebeckite-microgranite of Ailsa Craig, which travelled along the east coast of Ireland as far south as Cork and inland as far as Tyrone and Armagh. The link of this particular calcareous and shelly till-facies with a similar till on the east coast of Ireland has far-reaching significance with regard to the precise age of the deposits. In southern Ireland a till of this facies and derivation has been ascribed to the Munsterian Glacial period (Table 3). Further north, however, a shelly till at Glastry, Co. Down [J 640 630], known as the Glastry Lower Shelly Till, was derived from a south-west moving ice sheet. Shells in the till gave a C^{14} date of 24 050 $\pm$ 650 B.P. Dates from shells are notoriously unreliable but even so the field characteristics of the till suggests it is Midlandian age. A major ice movement towards the south and south-west in Co. Down in the Midlandian would be expected to have affected the whole north-east of Ireland and the most reasonable correlation seems to be with the Londonderry Lower Till. Until an interglacial episode is proved the problem will remain unresolved but for Northern Ireland the simplest solution at present is to consider all the known tills to be of last glaciation

(Midlandian) age and to be due to oscillations and the buffering of ice-sheets from a main Scottish source and a main Irish source.

The Scottish Ice (2) appears not to have over-ridden the Sperrins or central Donegal which probably continually held their own ice centres. It is likely that the problematical early till at Spincha Burn [C 738 043) at the head of the Roe Valley is equivalent in age to the Londonderry Lower Till but derived from a different but co-existing ice centre. With the growth of the Irish ice-sheet (3) the incursion of the Scottish ice-sheet was pushed back until it only marginally affected Antrim and Down, and was deflected so that in east Antrim the ice-flow was northwards. At the period of maximum of this glacial phase Ireland was covered by the ice-sheet as far south as the 'Tipperary Line', and mountains south of this had their own ice caps.

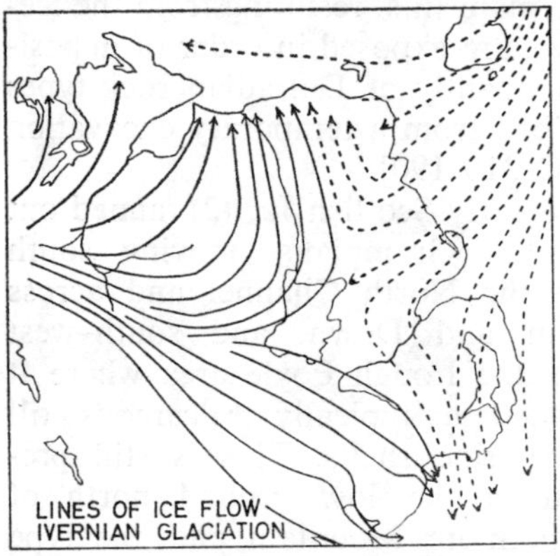

The centre of the Irish ice-sheet is not certain. It has been suggested that it was centred on Lough Neagh and that an ice-shed ran from Lough Neagh through Larne and up the Firth of Clyde, but recent work which claims to prove this on the basis of distribution of erratics is not universally accepted. In general it seems more likely that the main centre of Irish Ice was in the north–west, where high precipitation and high ground would provide an obvious focus.

The tills deposited from this last major Irish Ice movement (3) of late Midlandian age, are the most widespread. The till-facies vary widely, depending mainly on the underlying rock type. The close reflection of the underlying rock in the composition of the till is an interesting feature, with carry-over from one rock-type to another in direction of ice-movement being normally limited to quite small distances.

The story of continuing oscillations between the Scottish Ice and the Irish Ice is most easily proved where the south and west moving Scottish Ice actually impinged on northern Ireland, namely along the Co. Londonderry and Co. Down coasts. Inland, particularly in Fermanagh and Tyrone, where the ice from the Irish centre always dominated and movement was always in a similar direction, the task of distinguishing tills is much more difficult. However, several movements of the ice have been indicated by drumlin patterns and erratic carry in Co. Fermanagh. Here the most important recent discovery is a moss detritus found in a small freshwater basin of fine silts and sands within a drumlin section at Derryvree [H 361 390]. This gave a C^{14} age of $30\,000 \pm 1170/1030$ years B.P. indicating a Middle Midlandian age, possibly equivalent to the Brandon Interstadial of the Upton Warren Interstadial Complex in England. The insect fauna, mosses and flora indicate open tundra conditions at the time. A till underlying this interstadial deposit may be derived from an ice advance in the early Midlandian. The till fabric suggests a north–west source in the

Tertiary dolerite sill of Fair Head transgressing upward through Carboniferous rocks, with a lower non-transgressive sill near the bottom of the headland. The slopes of Murlough Bay in the foreground are all landslip. An extensive block scree can be seen below the cliff. Chalk and Tertiary lavas of Rathlin Island in the background.

PLATE 13

A. *Striated glacial pavement on Silurian greywacke. Carryduff Quarry.*

PLATE 14

B. *Corrie in Mourne granite. North face of Slieve Corragh.*

Donegal area and may correlate with the Lackagh Till of Co. Londonderry. A glacial advance at this period would be in accord with the Arctic conditions postulated to occur in north–west Europe at this time (Table 3).

TABLE 3

Relative ages of Quaternary deposits in Ulster

Stages		*A* *Hypothesis that no deposits remain from a pre-Midlandian glaciation*	*B* *Hypothesis that deposits of two glaciations are represented*
Midlandian (Ivernian)	Upper	Londonderry Upper Till (≡ Gelvin Till) Londonderry Lower Till (≡ Bovevagh Till and Glastry Lower Shelly Till, dated 24 050 B.P.) Lackagh Till (≡ Blanket Nook Till)	Londonderry Upper Till (≡ Gelvin Till)
	Middle	Derryvree Mud (dated 30 500 B.P.) Derryvree Till (? ≡ Lackagh Till and Blanket Nook Till)	Derryvree Mud
	Lower		
Ipswichian		Not known	Not known
Munsterian			Spincha Till Londonderry Lower Till (≡ Derryvree Till) Lackagh Till
Gortian		Not known	Not known

The dissolution of the last major ice-sheet was a gradual process with spasmodic advances and retreats of the ice front giving a series of moraines in the Carlingford area. One important halt or limited readvance gave the 'Carlingford Line' of marginal features from Carlingford to Galway.

The last ice-sheet to affect Ulster was the Scottish Ice (4) which, during the Antrim Coast Readvance, moved into north Antrim and left the Armoy-Ballymoney moraine to mark its limit. Opinion differs sharply on the extent to which this ice moved down the Irish Sea and west along the north

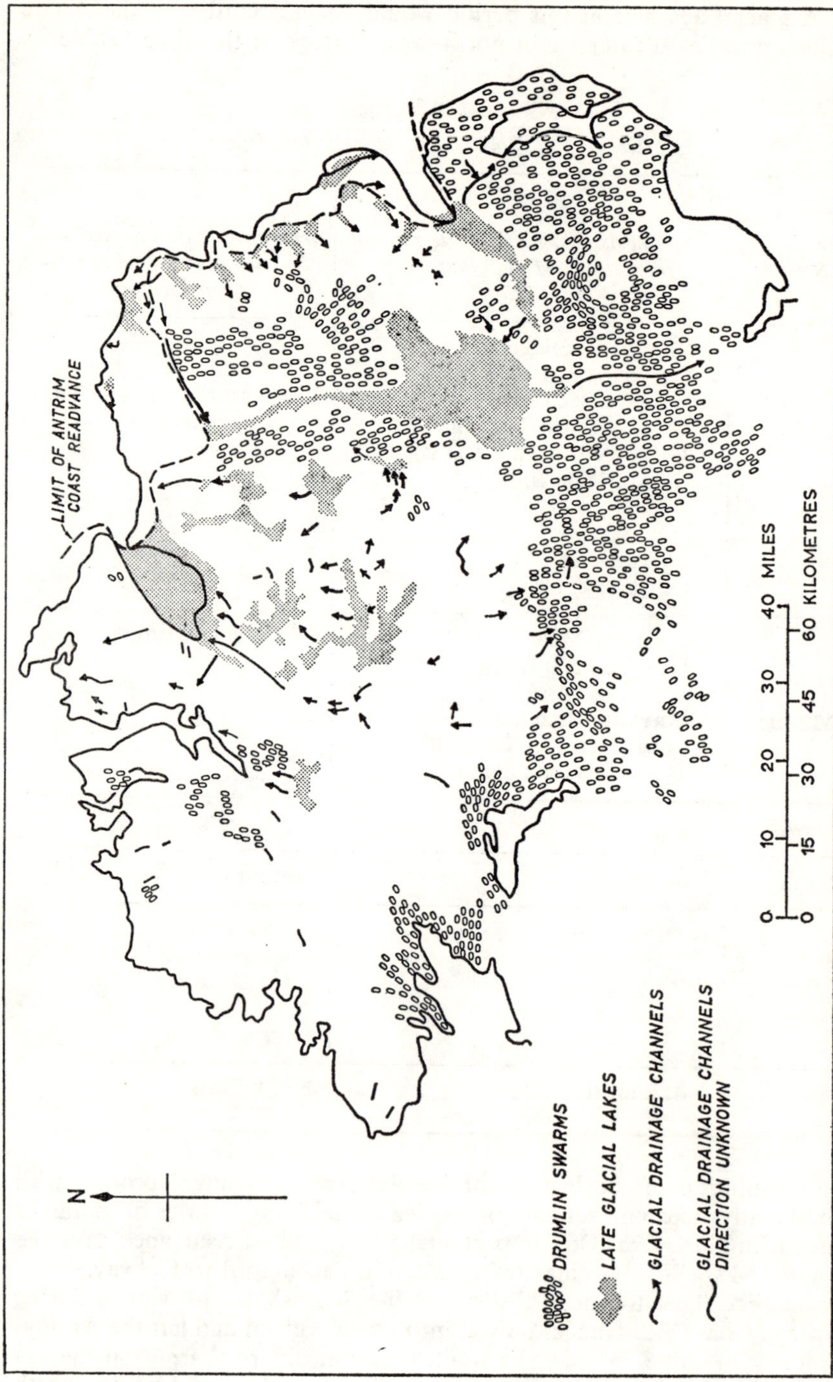

FIG. 33. *Distribution of drumlins, temporary glacial lakes and glacial drainage channels*

coast. The generally accepted view is that it extended as far as the mouth of the Lagan and the Foyle, a conclusion based largely on the evidence of glacial drainage channels and ice-ponded lakes. The ice-dammed Lake Lagan overflowed westwards through the Soldierstown gap towards Lough Neagh; the Bann drained southwards across the Scarva–Poyntzpass col to Newry; and a succession of lakes in the Antrim glens flowed north and west across inter-valley spurs to Ballycastle and thence along the coast to the west. The evidence for a westerly extension of the ice to the Foyle is based on the possibility that the Pennyburn Channel was used at this time to carry the Foyle drainage westwards, and on the occurrence of Scottish erratics in Donegal and on Inishtrahull.

Recent workers suggest that the readvance southwards to Belfast Lough cannot be substantiated and maintain that the absence of Scottish drift and the distribution of erratics from the Cushendun Granite indicate only north-moving ice in this area, while the presence of a late-glacial raised beach suggests that there was no ice at all in east Antrim during the re-advance period in the north. In the Lagan Valley, however, two boulder clays separated by the Malone Sands are significantly different in physical properties; the higher water content, plastic limit and liquid limit of the upper clay indicate much lower ice-loading. This upper clay may date from the Scottish Readvance, and it is not considered that its extent to the Lagan Valley is conclusively disproved.

Glacial deposits and phenomena are well displayed in many areas of Ulster. Of particular interest, apart from those mentioned above, are the boulder clay drumlins which are a striking feature of the landscape of north Down (Plate 15A). Rock basin lakes, *roches moutonnées* and ice striated surfaces are well developed on the top of Fair Head, near Ballycastle; ice striae are also seen on a glacially smoothed 'pavement' of basalt at Carnanee Quarry [C 823 355] near Portstewart, and at Carryduff Quarry [J 365 647], (Plate 13A). Glacial outwash sands with kettle holes are widely distributed, some of the most spectacular kettle holes occurring in the River Faughan valley at Ardlough [C 467 160] and from the grounds of Gransha Hospital to Templetown. The Loughs Enagh, Western and Eastern [C 468 195], occupy two particularly large kettle holes. Other extensive areas of outwash kames occur near Capecastle [D 085 365], around Pomeroy, in the country east of Ballygawley and in the upper valley of the Ballinderry River, west of Cookstown. Good examples of eskers are seen around Lisburn—e.g. at Causeway End [J 245 646], at Muntober [H 740 810] west of Cookstown, and at Eskermore [H 525 678] near Beragh. It is of interest that the terms 'drumlin' and 'esker' are Irish words which are now used internationally.

Glacial drainage channels can be readily seen at Bernisk Glen [H 605 675], Breen [D 125 335], Butterlope Glen [H 492 950], Altiffirnan Glen [D 148 365] and Loughaveema [D 207 354] where extensive lake deltas, formed in an ice-dammed lake, are seen north of the outfall. One particularly spectacular overflow channel cut deeply into solid Dalradian psammites and pelites is Fincarn Glen, near Londonderry City [C 478 157]. Another is the Gortin Gap [H 490 850], Co. Tyrone, where spectacular mounded sand and gravel has built into the Owenkillew valley. Other water-cut channels, which may have originated subglacially in some cases, are widespread e.g. on the north face of Carneighaneigh [D 184 373].

High level moraines and corries, relics of the last mountain-glacier phase, occur in the high ground of the Mourne Mountains (Plate 14B) and Donegal. Cirque-like hollows have also been identified on the northern slopes of Sawel Mountain above 380 metres but their age is uncertain.

Late-glacial and post-glacial events

From the appearance of the first tundra vegetation in the wake of the disappearing ice-sheets in late Midlandian times the climatic history of Ireland can be traced in the plant and tree pollen preserved in peat layers which occur at various levels in the post-glacial deposits, and also in the fossil mammals and molluscs. The relic of high-level marine and lake beaches and platforms indicate that there has also been considerable variation in the relative heights of sea and land since late-glacial times (Plate 16B). These events are summarized in the diagram (Table 5).

Some of the earliest absolute datings have been obtained from a site at Roddans Port, Co. Down. Deposits from the base of Pollen-Zone III have been dated by C^{14} to 11 500 B.P. Deposits belonging to Pollen-Zone II and I also occur at the site (Table 5) but the boundaries between them cannot be accurately defined at present.

The late-glacial period was cold with a tundra vegetation characterized by abundance of the dwarf willow *Salix herbacea* and by solifluxion deposits on hill slopes. Ice wedge pseudomorph structures, the fossil form of ground ice wedges, are known from fluvio-glacial sand and gravel deposits in the Sperrins and Co. Londonderry. It is interesting that the earliest record of ice-wedges in the British Isles was made in Co. Londonderry. Cryoturbation structures are commonly associated with them. A milder interval—the Allerød period—brought in the birch woods characteristic of later periods and is also of interest because of the widespread occurrence of the Giant Irish Deer *(Megaceros giganteus)* whose wide antlers have been found all over Ireland.

Post-glacial time saw a general rise in temperature with the concomitant arrival of hazel, alder, pine, elm and oak. A general rise in sea level of 20 m or more in Zone VI, the Flandrian transgression, inundated the coast and river valleys covering thin peats in the Lagan and Bann estuaries with grey unconsolidated sands, muds and silts. The peats in the Lagan estuary have been shown by radiocarbon dating to be 9100 ± 200 years old. These Estuarine Clays, which also occur in Larne Lough, Strangford Lough and Lough Foyle, contain an abundant fauna of molluscs and foraminifera. The latter, in the Belfast district, indicate that tidal flat conditions with *Elphidium excavatum* at the beginning of the transgression—wild boar *(Sus scrofa)* occurs just above the peat—were succeeded by low-salinity open water with *Ammonia beccarii*, and the highest beds show increasingly saline conditions with many species of foraminifera. Among the most abundant molluscs are the bivalves *Abra alba, Cerastoderma edule, Mytilus edulis, Ostrea edulis,*

TABLE 4

A curve of Weichselian climate change and a correlation of Pleistocene stage names in Ireland, Britain and Europe.

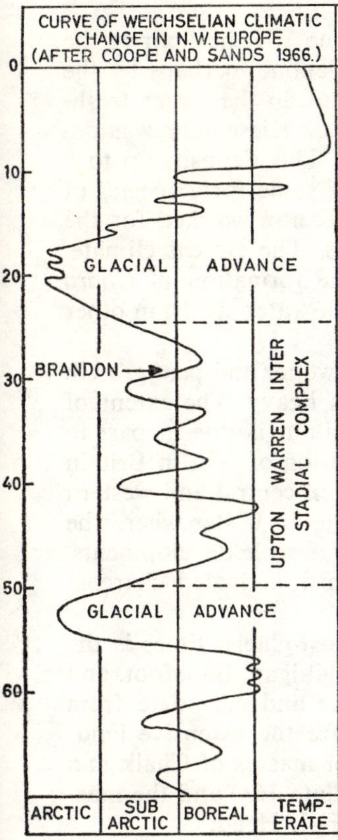

N.W. EUROPE STAGES	BRITISH STAGES	IRISH STAGES	IRISH SYNONYMS
FLANDRIAN	FLANDRIAN	LITTLETONIAN	POST-GLACIAL PERIOD
WEICHSELIAN	DEVENSIAN	MIDLANDIAN	MIDLAND GENERAL GLACIATION
EEMIAN	IPSWICHIAN	NOT KNOWN	—
SAALE	WOLSTONIAN	MUNSTERIAN	EASTERN GENERAL GLACIATION
HOLSTEIN	HOXNIAN	GORTIAN	—

Scrobicularia plana and *Venerupes decussata* and the gastropods *Bittium reticulatum, Littorina littorea,* and *Rissoa membranacea.*

The clays, known locally and graphically as 'sleech', are up to 17 m thick beneath Belfast and their very low bearing strength raises considerable engineering problems in the central area of the city where all the larger buildings rest on a forest of piles.

During the maximum transgression in Jessen's Zone VII the floor of the Bann Valley was flooded in the area south of Portglenone, perhaps by the ponding back of the drainage by higher sea level, and, in the larger fresh-water lake which preceded Lough Neagh, diatomite or Kieselguhr was deposited on top of the flooded peat of Zone VI age. This deposit, up to a metre or so thick, is formed of siliceous skeletons of some forty species of diatom which flourished in the warmer climate. It is now worked for the manufacture of silica brick and abrasives (Plate 17). The Boreal climate, probably 2·5° C warmer than today, encouraged the formation of *Chara* marls, white chemically-precipitated $CaCO_3$ with fresh-water shells, in other areas.

The recent period Zone VIII, has been cooler and wetter and peat growth on the hills—ombrogenous bog or blanket bog—was heavy. The extent of peat deposits in Ireland as compared with Great Britain is due in part to the wetter climate but also to more intensive exploitation of peat in Britain in the Middle Ages. Though very large areas of peat in central and western Ireland are commercially exploited there are few areas in Ulster where the volume of basin peat is sufficiently large to justify large scale developments, though it is worked for peat litter and the production of activated charcoal in counties Armagh and Londonderry.

The only other widespread deposit to date from post-glacial times is the blown sand which occupies extensive areas in the Magilligan, Bannfoot, and Dundrum Bay districts. Most of the dunes are stable and may date from Zone VII time. Also of late- and post-glacial age are the extensive land-slips around the Antrim plateau, usually consisting of masses of Chalk and basalt which have moved on the weak Lias shales (Plate 16A) and the massive block screes which are seen around the foot of Fair Head.

Human occupation of Ireland is known with certainty to date only from Mesolithic times, the earliest arrivals having crossed the land bridge from Scotland before the Flandrian transgression. The earliest known site is at Toome Bay where mesolithic flint implements occur below peat of Zone VI age, and other sites mainly of later age are known at Larne, Cushendun and the Bann Valley, associated with the raised-beach deposits of the maximum transgression.

Neolithic man left traces of his activity around the coasts of Antrim and Down, with kitchen middens and flint implements known at Dundrum, Rathlin, Ballycastle, White Park Bay, Portstewart and Castlerock. Particularly interesting are the axe factories at Tievebulliagh, near Cushendall, Co. Antrim, and Brockley on Rathlin Island, where tough hornfels was used to make heavy axes which have been found all over the British Isles. Late in the Neolithic came the first use of metals—bronze tools and weapons and gold ornaments—while the Megalith builders left their relics in the Court Cairns, Passage Graves, and stone circles which dot the Irish countryside.

The late Bronze Age, in Zone VIII, saw widespread deforestation as the agriculturalists spread and from the succeeding Iron Age date the ring forts or fortified farm-dwellings, and lake-dwellings or crannogs, the signs of a settled population.

References

Charlesworth 1924, 1939, 1955, 1963; Colhoun 1970, 1971; Dwerryhouse 1923; Hill and Prior 1968; Jessen 1949; Mitchell 1951; Morrison and Stephens 1965; Orme 1966; Smith 1961; Stephens 1957, 1963; Stephens and Synge 1966; Synge and Stephens 1966.

16. Mineral Deposits

Bauxite. The Interbasaltic Beds of Co. Antrim were worked for bauxite, originally as 'alum clay' for the manufacture of aluminium sulphate and later as an ore of aluminium. Work ceased in 1934 but was recommenced intensively during World War II when 240 000 tons were mined. Considerable reserves of both ferruginous bauxite (up to 30 per cent Fe_2O_3) and siliceous bauxite (up to 50 per cent SiO_2) remain.

Brick Clay. Apart from the Carboniferous clays, bricks and clay products are made from Keuper Marl, and from glacial clays at Aghadowey, Killough and Irvinestown.

Building Stone. The Triassic sandstones of Scrabo and the Lagan Valley and the Carboniferous sandstones of Tyrone (particularly near Dungannon), Londonderry, and Ballycastle were formerly worked for masonry but have long been abandoned. Mourne and Newry granites are still wrought for building and monumental purposes from several quarries in the Kilkeel and Newry areas, as are the Dalradian schists of Claudy, Co. Londonderry.

Coal. The only known coalfields in Northern Ireland are the small faulted areas of Namurian and Westphalian in Co. Tyrone, and the Viséan and Namurian area at Ballycastle. Thin coals are known from borehole evidence in the uppermost Viséan succession in the Foyle basin but are of no economic importance. It is possible also that the sedimentary basins indicated by geophysical evidence below the Antrim basalts at Larne and in the Ballycastle–Ballymoney area may contain deeply buried Coal Measures but this is as yet unproven.

In the Tyrone coalfield, Namurian coals occur in the area north–east of Dungannon and the workable seams, the Drumglass and Congo coals, have been worked out near the outcrop. These seams thin and split up eastwards and there is no prospect of economic working of the unwrought areas.

The Westphalian coals in Tyrone include thirteen seams in 275 m of mainly argillaceous strata. Eight seams have been worked but the area is much faulted and only small areas remain untouched.

The Ballycastle Coalfield, where the coals cropped out on coastal cliffs, was worked from an early date and production ceased in 1967. Only the Main Coal was of any real worth and it and the Hawk's Nest Coal have been worked out over virtually the whole field. There are fairly large reserves of low-quality Viséan coals beneath Fair Head.

Copper. Small shows of copper are recorded from Cappagh, Co. Tyrone, near Ballygawley, near Belleek, and at several localities in counties Down and Armagh. Small quantities of native copper have been found in the Antrim Lavas and high copper anomalies have been recorded for the geochemical reconnaissance of the Antrim plateau.

Diatomite. Extensive deposits of diatomite or 'Bann Clay' are worked between Toome and Portglenone on the alluvial flats of the Bann. The diatomite, 0·6 to 0·9 m in thickness, occurs beneath a peat cover. On drying and beneficiation it forms a fine white inert powder, used in manufacture of insulating refractory bricks.

Dolomite. Dolomitic limestone of Carboniferous age is known in the

A. *Drowned drumlins. Ringneill, Strangford Lough.*

PLATE 15

B. *Section through the Lisburn Esker, Lambeg. Horizontally-bedded gravels and sand with small slump features.*

A. *Late- or post-glacial landslips in Triassic, Liassic, Cretaceous and Tertiary rocks, below the Tertiary lava scarp. Binevenagh.*

PLATE 16

B. *Raised beach platform on the east side of the Mourne Mountains. Annalong.*

A. *Diatomite workings. The Creagh, Toomebridge.*

PLATE 17

B. *Diatomite working. Blocks of diatomataceous earth are cut by hand, and air dried by stacking in the open.*

Belleek area and at Kildress, Co. Tyrone. The Permian Magnesian Limestone of Cultra, Co. Down and Co. Tyrone is also a dolomite of commercial quality.

Feldspar. Pegmatite veins in the Moinian rocks near Castle Caldwell have been worked as a source of alkali feldspar for the pottery industry at Belleek, but working has long been abandoned. There are still some reserves, mainly in Co. Donegal.

Fireclay and Shale. Fireclay is worked in the Coalisland district for the manufacture of fireclay goods and for blending with other clays. Carboniferous shales and mudstones are worked on a large scale for brick and cement manufacture. Reserves are large but heavy overburden of glacial deposits makes extensive development difficult. The Lough Neagh Clays are worked in the Coalisland area as ball clays.

Flint. Hand-sorted flint from Chalk quarries is ground and used as poultry grit.

Glass-sand and moulding sands. Sandstones suitable for glass-making occur at Ballycastle, Kildress, Co. Tyrone, and in Co. Fermanagh. Moulding sands have been obtained from the glacial drift and from the Triassic sandstones of the Lagan Valley, and from some beach and dune areas.

Gold. Detrital gold has been found in the Moyola River, Draperstown, and the Glendun River, Co. Antrim. It is derived from quartz veins in the Dalradian schists but quantities are very small. Gold, associated with antimony and arsenical pyrite, is recorded from Monaghan where it occurs in a vein in Silurian sandstones.

Gypsum and anhydrite. Deposits of gypsum at Kingscourt, Co. Cavan, are worked on a large scale. In Northern Ireland a 5 m bed of anhydrite has been proved under Belfast but has not yet been exploited. Small quantities of gypsum were, at one time, hand-picked from the Keuper Marl in Belfast brickpits and used in cement manufacture.

Iron Ore. Extensive workings in the Interbasaltic iron ores of Antrim during the latter half of the nineteenth century, and the first two decades of the twentieth, removed some 5 million tons of ore. It is improbable that there is any substantial reserve near outcrop of the better class ores which were then worked. Though there are large quantities of low grade bauxitic ore with less than 30 per cent iron it is unlikely that it will be economic to work in the foreseeable future.

Blackband ironstone was worked in the Ballycastle Coalfield up to 1880. The seam passes laterally into coal to the east but there are large reserves below sea level to the west of the workings at Carrickmore. The seam was only about 0.6 m thick and further working is unlikely.

A bed of siliceous hematite 3 m thick occurs near the base of the Ordovician Volcanic Series in Co. Tyrone.

Bodies of hematite which have been explored as ore sources occur in a fault breccia in the Silurian rocks at Dechommet, Co. Down; in the Lower Carboniferous rocks near Cookstown; and in fault zones on Slieve Gallion.

Bog iron ore was extensively worked in Co. Londonderry at one time, and was used for gas purification. It is no longer dug.

Lead. Small veins of galena occur in many localities in the Silurian rocks of Down and Armagh, but only at Conlig, near Newtownards, have they ever proved of workable size. The Conlig vein occurs in crush-breccia and is associated with a Tertiary dyke. There are also traces of galena in the metamorphic rocks of the Sperrin Mountains.

Lignite. Thick beds of lignite in the Lough Neagh Clays have been worked as fuel on a small scale in the past. The lignites contain appreciable quantities of waxes. There are large reserves in the Thistleborough area, south–west of Crumlin.

Limestone. The Chalk and Carboniferous limestones are worked on a very large scale, mainly as a source of agricultural ground limestone, though large quantities of the former are also used in cement manufacture at Magheramorne and the latter for cement near Cookstown. Smaller quantities of Chalk are worked for the production of whiting, putty, etc. Limestones from both sources are burnt on a limited scale for quicklime.

Oil and natural gas. Though not yet proved in the sedimentary basins in the Irish Sea the extensive salt deposits of South Antrim indicate that suitable conditions for the occurrence of hydrocarbons may occur in the Antrim coastal areas. Small gas shows are recorded from bores in the Carboniferous rocks of Co. Fermanagh.

Oil-shale. Beds of low-grade oil shales occur in the Viséan succession in the Ballycastle Coalfield. Yields of up to 17.4 gal/ton were reported in 1917 but the shales are of no economic importance at the present day.

Peat. Small workings in hill and basin peat for domestic fuel are widespread throughout Northern Ireland. Bogs of sufficient extent and thickness for large-scale mechanized production are few, but peat-moss litter is produced south of Lough Neagh and activated charcoal is made from peat in Co. Londonderry.

Perlite. Perlitic obsidian (volcanic glass) occurs sporadically among the rhyolite lavas near Tardree. It has been worked on a small scale for the production of light-weight aggregate and insulating material, but most of the known occurrences are of poor quality.

Roadstone and concrete aggregate. Large quantities—some 12 million tons per annum—of basalt, dolerite, Carboniferous Limestone and Lower Palaeozoic grits are quarried for these purposes. A certain amount is exported but most is used locally. All sources give excellent material, provided adequate quality control is exercised on the quarry floor, with rigorous exclusion of vesicular and kaolinized basalt flow-tops and shaly partings in the grits. The Chalk is used as an aggregate for white concrete, valued for its architectural effect.

Some of the Lower Palaeozoic shales of Co. Down, and of Pomeroy, are exfoliated by heating to a temperature of 1300° and are suitable for the manufacture of light-weight aggregate.

Salt. Rock salt and brine have been worked at or near Carrickfergus for a century. Very large reserves of salt are available in the area from Carrickfergus to Larne and are worked for rock salt at Kilroot.

Sand and gravel. About 3 million tons per annum are worked, mainly from East Tyrone and Lough Neagh. Virtually all is derived from fluvioglacial deposits.

Water. Most of the public supplies in Northern Ireland have traditionally been from springs and surface catchments, supplemented in recent years by direct intake from rivers and lakes. Farm and domestic supplies have usually been from shallow wells in the drift deposits, or for short distances into solid rock.

In the late nineteenth century industry in Belfast started to use water from deep wells in the Triassic sandstones. The use of groundwater for public supply was pioneered in Lisburn and Newtownards which have for many years drawn much of their water from deep boreholes and, in the case of Newtownards, from a collecting tunnel beneath Scrabo Hill.

In recent years the development of the groundwater resources of the province has been accelerated and underground supplies are now being used by public bodies in many areas. The main aquifers and their potentials are summarized below.

Lower Palaeozoic	— very small supplies from shallow wells
Old Red Sandstone	— moderate yields of a few thousand gallons per hour from boreholes
Carboniferous Limestone	— very large yields from joint and fissure flow in suitable locations. Up to 35 000 g.p.h. known from deep boreholes
Carboniferous Sandstone	— unreliable, but little development has been done
Permo-Trias	— a major aquifer with yields of several thousand g.p.h. from boreholes
Cretaceous (Chalk)	— large yields locally from fissure flow, but this formation is usually drained naturally by springs
Tertiary Lavas	— large yields from joint flow in some favourable areas
Drift Deposits	— very large supplies are available from river gravel in suitable areas. Elsewhere domestic supplies from less permeable drift can be obtained

Zinc. Small quantities of zinc blende are recorded from some of the old lead workings of Co. Down.

References

Cameron and Sabine, 1969; Cameron, 1970; Cole 1912, 1922; Eyles 1952; Fowler 1959; Fowler and Robbie 1961; Wilson 1965; Wilson and Robbie 1966.

17. Bibliography

ADAMSON, J. H. and WILSON, G. F. 1933. The petrography of the Lower Carboniferous rocks of north-east Ireland. *Proc. Roy. Irish Acad.*, **41B**, 179–90.

AGRELL, S. O. and LANGLEY, J. M. 1958. The dolerite plug at Tievebulliagh, near Cushendall, Co. Antrim. *Proc. Roy. Irish Acad.*, **59B**, 93–127.

ANDERSON, J. G. C. 1948a. The occurrence of Moinian rocks in Ireland. *Quart. J. geol. Soc. Lond.*, **103**, 171–88.

—— 1948b. The stratigraphical nomenclature of Scottish metamorphic rocks. *Geol. Mag.*, **85**, 89–96.

ANDERSON, T. B. 1965. The evidence for the Southern Uplands Fault in north-east Ireland. *Geol. Mag.*, **102**, 383–92.

ANDREW, G. 1957. Old Red Sandstone of Portsalon. *Geol. Mag.*, **88**, 441.

ASHLEY, H. 1946. *Cheirotherium* footprint found at Scrabo Hill, Co. Down. *Irish Nat. J.*, **8**, 332.

BAILEY, E. B. 1959. Mobilization of granophyre in Eire and sinking of olivine in Greenland. *L'pool. Mnchr. geol. J.*, **2**, 143–54.

—— and MCCALLIEN, W. J. 1934. The metamorphic rocks of north-east Antrim. *Trans. Roy. Soc. Edinb.*, **58**, 163–77.

—— —— 1956. Composite minor intrusions and the Slieve Gullion Complex. *L'pool. Mnchr. geol. J.*, **1**, 466–501.

BARTON, R. 1757. *Lectures in Natural Philosophy, etc.* Dublin.

BERGER, J. F. 1816. On the geological features of the north-eastern counties of Ireland. With an introduction and remarks by W. Conybeare. *Trans. geol. Soc. Lond.*, (1) **3**, 121–222.

BISHOPP, D. W. 1951. The age of the Portsalon conglomerate. *Geol. Mag.*, **89**, 70.

BROWN, P. E. 1956. The Mourne Mountains granite, a further study. *Geol. Mag.*, **93**, 72–84.

CALDWELL, W. G. E. 1959. The Lower Carboniferous rocks of the Carrick-on-Shannon syncline. *Quart. J. geol. Soc. Lond.*, **115**, 163–186.

—— and CHARLESWORTH, H. A. K. 1962. Viséan Coral Reefs in the Bricklieve Mountains of Ireland. *Proc. Geol. Assoc.*, **73**, 359–82.

CAMERON, I. B. 1970. Sources of aggregate in Northern Ireland. *Rep. No.* 70/5, *Inst. geol. Sci.*

—— and SABINE, P. A. 1969. The Tertiary welded tuff vent agglomerate and associated rocks at Sandy Braes, Co. Antrim. *Rep. No.* 69/6, *Inst. geol. Sci.*

CHARLESWORTH, J. K. 1924. The glacial geology of the north-west of Ireland. *Proc. Roy. Irish Acad.*, **36B**, 174–314.

—— 1937. The olivine dolerite sill of The Maidens, Co. Antrim. *Irish Nat. J*, **6**, 265–73.

—— 1939. Some observations on the glaciation of north-east Ireland. *Proc. Roy. Irish Acad.*, **45B**, 255–95.

—— 1953. *The Geology of Ireland: an introduction.* Edinburgh.

—— 1955. The Carlingford re-advance between Dundalk, Co. Louth, and Kingscourt and Lough Ramore, Co. Cavan. *Irish Nat. J.*, **11**, 299–302.

—— 1963a. Some observation on the Irish Pleistocene. *Proc. Roy. Irish Acad.*, **62B**, 295–322.

—— 1963b. *Historical Geology of Ireland*, Edinburgh.

—— (Edit.). 1960. The geology of north-east Ireland. *Proc. Geol. Assoc.*, **71**, 429–60.

—— and ERDTMAN, G. 1935. Post-glacial section at Milewater Dock, Belfast. *Irish Nat. J.*, **5**, 234–5.

—— and HARTLEY, J. J. 1935. The Tardree and Hillsborough dyke swarms. *Irish Nat. J.*, **5**, 193–6.

—— and PRESTON, J. 1960. *Geology around the University Towns: North-East Ireland, the Belfast Area.* Geol. Assoc. Guide No. 18.

COBBING, E. J. C. 1964. The Highland Boundary Fault in east Tyrone. *Geol. Mag.*, **101,** 496–561.

—— 1969. Schistosity and folding in a banded gabbro from Tyrone. *Bull. Geol. Surv. Gt Br.*, No. 30, 89–97.

—— MANNING, P. I. and GRIFFITH, A. E. 1965. Ordovician-Dalradian unconformity in Tyrone. *Nature* **206,** 1132–50.

COLE, C. A. J. 1922. Memoir and Map of localities of minerals of economic importance and metalliferous mines in Ireland. *Mem. geol. Surv.*

—— (Edit.). 1912. *The Interbasaltic Rocks of north-east Ireland. Mem. geol. Surv.*

—— and HALLISSY, T. 1924. *Handbook of the Geology of Ireland.* London.

COLHOUN, E. A. 1970a. On the nature of the glaciations and final deglaciation of the Sperrin Mountains and adjacent areas in the north of Ireland. *Irish Geogr.* **6,** 162–185.

—— 1970b. Early record and interpretation of ice-wedge pseudomorph in County Londonderry, Northern Ireland, by J. R. Kilroe. *J. Glaciology* **9,** 391–2.

—— 1971a. The glacial stratigraphy of the Sperrin Mountains and its relation to the glacial stratigraphy of north-west Ireland. *Proc. Roy. Irish Acad.* **71B,** 37–52.

—— 1971b. Late Weichselian periglacial phenomena of the Sperrin Mountains, Northern Ireland. *Proc. Roy. Irish Acad.* **71B,** 53–71.

CONYBEARE, W. and BUCKLAND, W. 1816. Descriptive notes referring to the outline of sections presented by part of the coasts of Antrim and Derry. *Trans. geol. Soc. Lond.*, (1), **3,** 196–216.

COOPE, G. R. and SANDS, C. H. S. 1966. Insect faunas of the last glaciation from the Tame Valley, Warwickshire. *Proc. R. Soc., Lond.* B, **165,** 389–412.

DAWSON, J. 1951. The Brockley dolerite plug and Church Bay volcanic vent, Rathlin Island. *Irish Nat. J.*, **10,** 156–61.

DWERRYHOUSE, A. R. 1923. The glaciation of north-east Ireland. *Quart. J. geol Soc. Lond.* **79,** 352–421.

ELWELL, R. W. D. 1958. Granophyre and hybrid pipes in a dolerite layer of Slieve Gullion, *J. Geol.*, **66,** 57–71.

EMELEUS, C. H. 1955. The granites of the western Mourne Mountains, Co. Down. *Sci. Proc. Roy. Dub. Soc.*, **27,** 35–50.

—— 1962. The porphyritic felsite of the Tertiary ring complex of Slieve Gullion, Co. Armagh. *Proc. Roy. Irish Acad.*, **62B,** 55–76.

—— and PRESTON, J. 1969. *Field Excursion Guide to the Tertiary Volcanic Rocks of Ireland.* Belfast.

EVANS, E. E. (Edit.). 1952. *Belfast in its Regional Setting.* Brit. Assoc.

EVANS, W. B. and others, 1968: Geology of the country around Macclesfield. *Mem. Geol. Surv. G. B.*

EYLES, V. I. 1952. The composition and origin of the Antrim laterites and bauxites. *Mem. geol. Surv. N.I.*

FEARNSIDES, W. G., ELLES, G. L. and SMITH, B. 1907. The Lower Palaeozoic rocks of Pomeroy. *Proc. Roy. Irish Acad.*, **26B,** 97–128.

FLETCHER, T. P. 1967. Correlation of the Cretaceous exposures of east Antrim. *Unpublished M.Sc. Thesis, Queen's University, Belfast.*

FOWLER, A. 1955a. The zonal sequence in the Carboniferous rocks of south-west Tyrone. *Bull. Geol. Surv. Gt Br.*, No. 8, 38–43.

—— 1955b. The Permian of Grange, Co. Tyrone. *Bull. Geol. Surv. Gt Br.*, No. 8, 44–53.

—— 1959. The non-ferrous minerals of Northern Ireland: in *The future of non-ferrous mining in Gt. Britain and Ireland,* 27–34. *Inst. Min. Metall.*

—— and Robbie, J. A. 1961. Geology of the Country around Dungannon. *Mem. Geol. Surv. N.I.*

George, T. N. 1953. The Lower Carboniferous rocks of north-western Ireland. *Adv. Sci.*, 65–73.

—— 1958. Lower Carboniferous palaeogeography of the British Isles. *Proc. Yorks. geol. Soc.*, **31,** 227–318.

—— 1960. The stratigraphical evolution of the Midland Valley of Scotland. *Trans. geol. Soc. Glasg.*, **24,** 32–107.

—— 1967. Landform and structure in Ulster. *Scot. J. Geol.*, **3,** 413–8.

—— and Oswald, D. H. 1957. The Carboniferous rocks of the Donegal syncline. *Quart. J. geol. Soc. Lond.*, **113,** 137–79.

Goldring, D. C. 1956. The structural petrology of the Dalradian rocks of north-east Antrim. *Adv. Sci.*, **12,** 576.

—— 1961. The relationship of the microfabric to the small-scale structures of the Dalradian rocks of north-eastern Co. Antrim. *Proc. Roy. Irish Acad.*, **61B,** 345–68.

Griffith, A. E., Manning, P. I., and Wilson, H. E. The Geology of the Country around Carrickfergus and Bangor. *Mem. geol. Surv. N.I.* (In preparation).

Hancock, J. M. 1961. The Cretaceous system in Northern Ireland. *Quart. J. geol. Soc. Lond.*, **117,** 11–36.

Harper, J. C. and Hartley, J. J. 1937. A recently discovered Ordovician inlier in Co. Down. *Irish Nat. J.*, **6,** 253–5.

—— —— 1938. The Silurian inlier of Lisbellaw, Co. Fermanagh. *Proc. Roy. Irish Acad.*, **45B,** 73–87.

Harris, N. 1937. A petrological study of the Portrush sill and its veins. *Proc. Roy. Irish Acad.*, **43B,** 95–134.

Hartley, J. J. 1933. The geology of north-eastern Tyrone and the adjacent portions of County Londonderry. *Proc. Roy. Irish Acad.*, **41B,** 218–85.

—— 1936a. On the occurrences of *Saccamminopsis fusulinaformis* in Northern Ireland. *Irish Nat. J.*, **6,** 95–7.

—— 1936b. On the age of the igneous series of Slieve Gullion, Northern Ireland. *Geol. Mag.*, **73,** 226–8.

—— 1938. The Dalradian rocks of the Sperrin Mountains and adjacent areas in Northern Ireland. *Proc. Roy. Irish Acad.*, **44B,** 141–71.

—— 1940. The subsoils of Belfast and district. Inst. Civil Engineers (N. Ireland).

—— 1943. Notes on the Lower Marls of the Lagan Valley. *Irish Nat. J.*, **8,** 128–32.

—— 1949. Further notes on the Permo-Triassic rocks of Northern Ireland. *Irish Nat. J.*, **9,** 314–6.

Hill, A. R. and Prior, D. B. 1968. Directions of ice-movement in north-east Ireland. *Proc. Roy. Irish Acad.*, **66B,** 71–84.

Hospers, J. and Charlesworth, H. A. K. 1954. The natural permanent magnetism of the Lower Basalts of Northern Ireland. *Mon. Not. Roy. Astro. Soc. Geophy Supp.*, **7,** 32–43.

Howard, A. J. 1956. *Northern Ireland Peat Bog Survey.* Final report of the preliminary survey. Belfast.

Hull, E. 1878. *Physical Geology and Geography of Ireland.* London.

Hume, W. F. 1897. The Cretaceous strata of County Antrim. *Quart. J. geol. Soc. Lond.*, 53, 540–606.

Jackson, J. S. 1965. The Upper Carboniferous (Namurian and Westphalian) of Kingscourt, Ireland. *Sci. Proc. Roy. Dub. Soc.*, **21A,** 131–52.

Jessen, K. 1949. Studies in late-Quaternary deposits and flora-history of Ireland. *Proc. Roy. Irish Acad.*, **52B,** 85–290.

Johnstone, G. S. 1966. *British Regional Geology, The Grampian Highlands.*

JOHNSON, M. R. W. and STEWART, F. H. 1963. *The British Caledonides.* Edinburgh.

KING, W. 1853. On the Permian fossils of Cultra. *Rep. Brit. Assoc.* for 1852, 53.

—— 1857. On the occurrence of Permian Magnesian Limestone at Tullyconnel, near Ardtea, in the county of Tyrone. *J. Geol. Soc. Dublin,* **7,** 67–81.

LAMPLUGH, G. W. (Edit.). 1904. Geology of the country around Belfast. *Mem. geol. Surv.*

LANGTRY, G. 1874. On the occurrence of Middle Lias at Ballycastle. *Rep. Brit. Assoc.,* 88.

LE BAS, M. J. 1960. The petrology of the layered basic rocks of the Carlingford Complex, Co. Louth. *Trans. Roy. Soc. Edinb.,* **64,** 169–200.

—— 1967. On the origin of the Tertiary granophyres of the Carlingford Complex, Ireland. *Proc. Roy. Irish Acad.,* **65B,** 325–38.

MCCALLIAN, W. 1930. The gneiss of Inishtrahull. *Geol. Mag.,* **57,** 542–9.

—— 1931. A contribution to the correlation of the Dalradian rocks of Scotland and Ireland. *Geol. Mag.,* **68,** 126–33.

—— 1935. The metamorphic rocks of Inishowen. *Proc. Roy. Irish Acad.,* **42B,** 407–42.

—— 1936. A note on the Dalradian pillow lavas of Strabane. *Proc. Roy. Irish Acad.,* **43B,** 13–22.

MCGUGAN, A. 1957. Upper Cretaceous foraminifera from Northern Ireland. *J. Paleont,* **31,** 329–48.

MCKERROW, W. S. 1959. The Southern Upland Fault in Ireland. *Geol. Mag.,* **96,** 347–52.

MANNING, P. I., ROBBIE, J. A. and WILSON, H. E. 1970. Geology of Belfast and the Lagan Valley. *Mem. geol. Surv. N.I.*

MITCHELL, G. F. 1951. Studies in Irish Quaternary Deposits. *Proc. Roy. Irish Acad.,* **53B,** 111–206.

MORRISON, M. E. S. and STEPHENS, N. 1965. A submerged late-Quaternary deposit at Roddans Port on the north-east coast of Ireland. *Phil. Trans. Roy. Soc. Lond.,* B, **249,** 221–55.

NOCKOLDS, S. R. 1935. Contributions to the petrology of Barnavave, Carlingford, I.F.S. 1. The Junction Hybrids. *Geol. Mag.,* **72,** 289–315.

—— 1937. 2. An occurrence of quartz-bearing syenite and its xenoliths. *Geol. Mag.,* **74,** 125–32.

—— 1938. 3. On some hybrids from the east and south-east slopes of Barnavave Mountain. *Geol. Mag.,* **75,** 469–79.

NOLAN, J. 1880. The Old Red Sandstone of the north of Ireland. *Quart. J. geol Soc. Lond.,* **36,** 529–35.

OLD, R. A. 1970. A Tertiary agglomerate vent, Spencestown, Ballymena, Co. Antrim. *Irish Nat. J.,* **16,** 309–11.

ORME, A. R. 1966. Quaternary changes in sea-level in Ireland. *Trans. Inst. Brit. Geogr.,* **39,** 127–40.

OSWALD, D. H. 1955. The Carboniferous rocks between the Ox Mountains and Donegal Bay. *Quart. J. geol. Soc. Lond.,* **111,** 167–86.

PADGET, P. 1951. The geology of the Clogher-Slieve Beagh area, Co. Tyrone. *Sci. Proc. Roy. Dub. Soc.,* **26,** 63–83.

—— 1952. The Carboniferous beds at Kildress. *Irish Nat. J.,* **10,** 238–42.

—— 1953. The Stratigraphy of Cuilcagh, Ireland. *Geol. Mag.,* **90,** 17–26.

PATTERSON, E. M. 1946. The olivine-gabbro of Ardtrea, Co. Tyrone. *Irish Nat. J.,* **8,** 171–6.

—— 1950. Evidence of fissure eruption in the Tertiary lava plateau of north-east Ireland. *Geol. Mag.,* **87,** 45–52.

—— 1951a. An occurrence of quartz-trachyte among the Tertiary basalt lavas of north-east Ireland. *Proc. Roy. Irish Acad.,* **53B,** 265–87.

—— 1951b. Notes on three Tertiary dolerites from counties Down and Antrim. *Irish Nat. J.*, **10,** 177–81.

—— 1951c. A petrochemical study of the Tertiary lavas of north-east Ireland. *Geochim. Cosmochim. Acta*, **2,** 283–99.

—— 1955a. The Tertiary lava succession in the northern part of the Antrim plateau. *Proc. Roy. Irish Acad.*, **57B,** 79–122.

—— 1955b. The Tertiary lava succession in the western part of the Antrim plateau. *Proc. Roy. Irish Acad.*, **57B,** 155–78.

—— 1963. The Tertiary vents in the northern part of the Antrim plateau, Ireland. *Quart. J. geol. Soc. Lond.*, **119,** 419–43.

—— and SWAINE, D. J. 1955. A petrochemical study of Tertiary tholeiitic basalts: the Middle Lavas of the Antrim plateau. *Geochim. Cosmochim. Acta*, **8,** 178–81.

—— —— 1957. The Tertiary dolerite plugs of north-east Ireland. A survey of their geology and geochemistry. *Trans. Roy. Soc. Edinb.*, **63,** 317–31.

PITCHER, W. S., ELWELL, R. W. D., TOZER, C. F. and CAMBRAY, F. W. 1964. The Lennan Fault. *Quart. J. geol. Soc. Lond.*, **120,** 241–74.

—— and READ, H. H. 1959. The main Donegal granite. *Quart. J. geol. Soc. Lond.*, **114,** 259–300.

—— —— 1960. The aureole of the main Donegal granite. *Quart. J. geol. Soc. Lond.*, **116,** 1–36.

—— and SHACKLETON, R. M. 1966. On the correlation of certain Lower Dalradian successions in north-west Donegal. *Geol. J.*, **5,** 149–156.

—— —— and WOOD, R. S. R. 1971. The Ballybofey Anticline: a solution of the general structure of parts of Donegal and Tyrone. *Geol. J.*, **7,** 321–8.

POLLOCK, J. and WILSON, H. E. 1961. A new fossiliferous locality in Co. Down. *Irish Nat. J.*, **13,** 244–8.

PORTLOCK, J. E. 1843. *Report on the geology of the county of Londonderry and parts of Tyrone and Fermanagh.* Dublin.

PRESTON, J. 1962. Explosive volcanic activity in the Triassic sandstone of Scrabo Hill, Co. Down. *Irish Nat. J.*, **14,** 45–51.

—— 1963. The dolerite plug at Slemish, Co. Antrim, Ireland. *Lpool Manchr. geol. J.*, **3,** 301–14.

—— 1965. Tertiary feeder dykes in the west of Ireland. *Proc. geol. Soc. Lond.*, 1626, 149–50.

—— 1967a. The Blind Rock dyke, Co. Donegal. *Irish Nat. J.*, **15,** 286–93.

—— 1967b. A Tertiary feeder dyke in Co. Fermanagh, Northern Ireland. *Sci. Proc. Roy. Dub. Soc.*, **3A,** 1–16.

RAST, N. and LITHERLAND, M. 1970. The correlation of the Ballachulish and Perthshire (Iltay) Dalradian successions. *Geol. Mag.*, **107,** 259–72.

REID, R. E. H. 1958. Remarks on the Upper Cretaceous Hexactinellida of County Antrim. *Irish Nat. J.*, **12,** 236–43.

—— 1959. Age of the Cretaceous basal conglomerate at Murlough Bay, Co. Antrim. *Geol. Mag.*, **96,** 86–7.

—— 1962. The Cretaceous succession in the area between Red Bay and Garron Point, Co. Antrim. *Irish Nat. J.*, **14,** 73–7.

—— 1963. New records of *Gonioteuthis* in Ireland. *Irish Nat. J.*, **14,** 98.

—— 1964. The Lower (pre-*Belemnitella mucronata*) White Limestone of the east and north-east of Co. Antrim. *Irish Nat. J.*, **14,** 262–9, 296–303.

—— 1971. The Cretaceous rocks of north-eastern Ireland. *Irish Nat. J.*, **17,** 105–29.

REYNOLDS, D. L. 1928. The petrography of the Triassic sandstones of north-east Ireland. *Geol. Mag.*, **65,** 448–73.

—— 1931. The dykes of the Ards peninsula. *Geol. Mag.*, **68,** 97–111, 145–65.

—— 1934. The eastern end of the Newry Igneous Complex. *Quart. J. geol. Soc. Lond.*, **90,** 585–636.

—— 1936. The two monzonitic series of the Newry complex. *Geol. Mag.*, **73,** 337–64.

—— 1936. The augite-biotite-diorite of Newry complex. *Geol. Mag.*, **73,** 560–2.

—— 1943a. Granitization of hornfelsed sediments in the Newry granodiorite of Goraghwood Quarry, Co. Armagh. *Proc. Roy. Irish Acad.*, **48B,** 231–67.

—— 1943b. The south-western end of the Newry igneous complex. *Quart. J. geol. Soc. Lond.*, 99, 205–46.

—— 1943c. The albite-schists of Antrim. *Proc. Roy. Irish Acad.*, **48B,** 43–66.

—— 1951. The geology of Slieve Gullion, Foughill and Carrickcarnan, etc. *Trans. Roy. Soc. Edinb.*, **62,** 85–143.

—— 1961. Lapies and solution pits in olivine-dolerite sills at Slieve Gullion, Northern Ireland. *J. Geol.* **69,** 110–7.

RICHEY, J. E. 1927. The structural relations of the Mourne Granites. *Quart. J. geol. Soc. Lond.*, **83,** 653–88.

—— 1932. The Tertiary ring complex of Slieve Gullion. *Quart. J. geol. Soc. Lond.*, **88,** 776–849.

—— 1935. Further evidence concerning the age of the gabbros of the Slieve Gullion district. *Proc. Geol. Assoc.*, **46,** 487–92.

—— 1940. Association of explosive brecciation and plutonic intrusion in the British Tertiary Igneous Province. *Bull. Vulcan* (2), **6,** 157–75.

ROBBIE, J. A. 1955a. The Slieve Binnian tunnel, an aqueduct in the Mourne Mountains, Co. Down. *Bull. Geol. Surv. Gt Br.*, No. 8, 1–20.

—— 1955b. The Carboniferous rocks of Edendork, Co. Tyrone. ibid, 21–37.

SABINE, P. A. 1968. Ferrian chlorospinel from Carneal, Co. Antrim. *Min. Mag.*, **36,** 948–54.

SCOULER, J. 1837. Observations on the lignite and silicified woods of Lough Neagh. *J. geol. Soc. Dublin*, **1,** 231–41.

SHARPE, E. N. 1970. An occurrence of pillow lavas in the Ordovician of County Down. *Irish Nat. J.* **16,** 299–301.

SHERLOCK, R. L. 1926. A correlation of the British Permo-Triassic Rocks. Part 1. *Proc. Geol. Assoc.*, **37,** 1–69.

—— 1928. A correlation of the British Permo-Triassic Rocks. Part 2. *Proc. Geol. Assoc.*, **39,** 49–95.

SIMPSON, I. M. 1955. The Lower Carboniferous stratigraphy of the Omagh syncline, Northern Ireland. *Quart. J. geol. Soc. Lond.*, **110,** 391–408.

SINGH, G. 1970. Late-glacial vegetational history of Lecale, Co. Down, Northern Ireland. *Proc. Roy. Irish Acad.*, **69B,** 189–216.

—— and SMITH, A. G. 1966. The post-glacial marine transgression in N. Ireland—conclusions from estuarine and raised beach deposits: a contrast. *Palaeobotanist*, **15,** 230–4.

SKIBA, W. 1952. The contact phenomena of the north-west of the Crossdoney Complex. *Trans. Edinb. geol. Soc.*, **15,** 322–45.

SMITH, A. G. 1961. Cannons Lough, Kilrea, Co. Derry: Stratigraphy and pollen analysis. *Proc. Roy. Irish Acad.*, **61B,** 369–83.

STEPHENS, N. 1957. Some observations on the 'Inter-glacial' platform and the early post-glacial raised beach on the east coast of Ireland. *Proc. Roy. Irish Acad.*, **58B,** 129–49.

—— 1963. Late-glacial sea levels in north-east Ireland. *Irish Geogr.*, **4,** 345–59.

—— and COLLINS, A. E. P. 1960. The quaternary deposits at Ringneill Quay and Ardmillan, Co. Down. *Proc. Roy. Irish Acad.*, **61C,** 41–77.

—— and SYNGE, F. M. 1966. Pleistocene shorelines: in *Essays in Geomorphology:* G. H. Drury, Edit. London.

STUBBLEFIELD, C. J. (in discuss.), DUNHAM, K. C. and ROSE, W. C. C. 1949. Permo-Triassic Geology of South Cumberland and Furness. *Proc. Geol. Assoc.*, **60,** 11–40.

SUESS, E. 1904. *The Face of the Earth,* London.

SWANSTON, W. S. and LAPWORTH, C. 1877. On the Silurian rocks of County Down, *Proc. Belfast Nat. Field Club.* App. **4,** 1876–7, 107–48.

SYNGE, F. M. and STEPHENS, N. 1966. Late and post-glacial shorelines and ice limits in Argyll and north-east Ulster. *Trans. Inst. Brit. Geogr.*, **39,** 101–125.

TATE, R. 1864. On the Liassic strata of the neighbourhood of Belfast. *Quart. J. geol. Soc. Lond.*, **20,** 103–11.

—— 1865. On the correlations of the Cretaceous formations of the north-east of Ireland. *Quart. J. geol. Soc. Lond.*, **21,** 15–44.

—— 1868. On the Lower Lias of the north-east of Ireland. *Quart. J. geol. Soc. Lond.*, **23,** 297–314.

—— 1870. On the Middle Lias in the north-east of Ireland. *Quart. J. geol. Soc. Lond.*, **26,** 324–5.

TILLEY, C. E. 1929. On Larnite and its associated minerals from the limestone contact-zones of Scawt Hill, Co. Antrim. *Min. Mag.*, **22,** 77–86 (See also *ibid.* **23, 24, 25, 26** for subsequent papers).

—— and HARWOOD, H. F. 1931. The Dolerite-Chalk contact of Scawt Hill, Co. Antrim. The production of basic alkali-rocks by the assimilation of limestone by basaltic magma. *Miner. Mag.* **22,** 439–68.

TOMKEIEFF, S. I. 1934. Differentiation in basalt lavas, Island Magee, Co. Antrim. *Geol. Mag.* **71,** 501–12.

—— 1935. The dolerite plug at Ballygally Head, Co. Antrim. *Irish Nat. J.*, **5,** 252–7.

—— 1940a. The dolerite plugs of Tieveragh and Tievebulliagh near Cushendall, Co. Antrim, with a note on Buchite. *Geol. Mag.*, **77,** 54–64.

—— 1940b. The basalt lavas of the Giant's Causeway district, Northern Ireland. *Bull. Vulcanologique*, 2, **6,** 89–143.

—— 1943. The Carboniferous lavas of the Ballycastle district. *Irish Nat. J.*, **8,** 67–72.

—— and MARSHALL, C. E. 1935. The Mourne dyke swarms. *Quart. J. geol. Soc. Lond.*, **91,** 251–92.

—— —— 1940. The Killough-Ardglass dyke swarm. *Quart. J. geol. Soc. Lond.*, **96,** 321–38.

—— and PATTERSON, E. M. 1947. The Tertiary igneous rocks in the neighbourhood of Ballintoy, I. The Knocksoghey Sill. *Irish Nat. J.*, **9,** 89–96.

1948. II. The Carrickarade Volcano. *Irish Nat. J.*, **9,** 203–12.

—— —— 1953. III. The coast near Ballintoy Harbour. *Irish Nat. J.*, **11,** 35–45.

TURNER, J. S. 1952. The Lower Carboniferous rocks of Ireland. *L'pool. Manchr. geol. J.*, **1,** 113–47.

WALKER, G. P. L. 1951. The amygdale minerals in the Tertiary lavas of Ireland. I. *Min. Mag.*, **29,** 773–91; 1959, II. *Min. Mag.* **32,** 202–17: 1960. III. *Min. Mag.* **32,** 503–27.

—— 1959. Some observations on the Antrim basalts and associated dolerite intrusions. *Proc. Geol. Assoc.*, **70,** 179–205.

—— 1960. An occurrence of mugearite in Antrim. *Geol. Mag.*, **97,** 62–4.

—— 1962. A note on occurrences of tree remains within the Antrim basalts. *Proc. Geol. Assoc.*, **73,** 1–7.

—— and LEEDAL, G. P. 1954. The Barnesmore granite complex. *Sci. Proc. Roy. Dub. Soc.*, **26,** 207–43.

WATTS, W. A. 1962. Early Tertiary pollen deposits in Ireland. *Nature*, **193,** 600.

—— 1963. Fossil seeds from the Lough Neagh Clay. *Irish Nat. J.*, **14,** 117–8.

WHITEHURST, J. 1786. *On the Original State and Formation of the Earth.* London.

WILKINSON, S. B., MCHENRY, A., KILROE, J. R. and SEYMOUR, H. J. 1908. The Geology of the country around Londonderry. *Mem. geol. Surv.*

WILSON, H. E. 1953. The petrography of the Old Red Sandstone rocks of the north of Ireland. *Proc. Roy. Irish Acad.*, **55B,** 283–320.

—— 1965. The rise and decline of the iron ore and bauxite industry of Co. Antrim. *Proc. Belfast Nat. Hist. and Phil. Soc.*, II, **7,** 14–23.

—— and ROBBIE, J. A. 1966. The Geology of the country around Ballycastle. *Mem. geol. Surv. N.I.*

WILSON, R. L. 1964. The tertiary dykes of Magho Mountain, Co. Fermanagh. *Irish Nat. J.*, **14,** 254–57.

WOOD, C. J. 1967. Some new observations on the Maestrichtian Stage in the British Isles. *Bull. geol. Surv. Gt Br.*, *No.* 27, 271–88.

WRIGHT, W. B. 1919. An analysis of the Palaeozoic floor of north-east Ireland. *Sci. Proc. Roy. Dub. Soc.*, **15,** 629–50.

—— 1924a. Geology of the Ballycastle Coalfield. *Mem. geol. Surv. Ireland.*

—— 1924b. Age and origin of the Lough Neagh Clays. *Quart. J. geol. Soc. Lond.*, **80,** 468–88.

—— and MUFF, H. B. 1904. The Pre-glacial Raised Beach of the South Coast of Ireland. *Sci. Proc. Roy. Dub. Soc.*, **10,** n–s, 250–308.

YATES, P. J. 1956. Upper *Eumorphoceras* Beds on Benbrack, Co. Cavan. *Geol. Mag.*, **93,** 85.

—— 1962. The palaeontology of the Namurian rocks of Slieve Anierin, Co. Letrim, Eire. *Palaeontology*, **5,** 355–443.

18. Appendix

Geological Itineraries

It is possible to see many of the most interesting geological features in Ulster in a few days and for the visitor with only a short period at his disposal the following routes are suggested. Maps referred to are Ordnance Survey Third series one-inch to one mile maps and Geological Survey one-inch sheets. Publications which will be found useful are:

Geologists Association Guide No. 18: Geology around the University towns: North–East Ireland—the Belfast area. The Scientific Anglian, 30/30A St. Benedicts' St., Norwich NOR 24J. Price 20p.

Field Excursion Guide: Tertiary Volcanic Rocks of Ireland. Emeleus and Preston. University Bookshop, University Road, Belfast. Price 25p.

The Antrim Coast Road, with notes and sketches on its geology and scenery. P. S. Rhodes. Ulster Tourist Development Association, High Street, Belfast. Price 10p.

A. Metamorphic Rocks

1. *North Antrim*

   Belfast, Ballymena, Cushendun, Torr Head, Murlough Bay, Ballycastle. Rugged and difficult cliff and coastal sections.
   References: Bailey and McCallien 1934; Goldring 1956, 1961; Wilson and Robbie 1966. [O.S. 1, 3: G.S. 8 14]. About 190 km.

2. *Tyrone*

   Belfast, Cookstown, minor roads to Oritor [H 774 796], Corvanaghan [H 718 812], Lough Fea, Glenelly Valley, Donemana, Strabane, Newtownstewart, Pettigo, Laghy, and minor roads to Unshin [G 941 633]. Return via Belleek.
   References: Anderson 1948a; Hartley 1933, 1938; McCallien 1936. [O.S. 2, 4, 5: G.S. 17, 25, 26, 32, 33]. About 440 km.

B. Lower Palaeozoic Rocks

1. *Co. Down*

   Belfast, Helen's Bay, Donaghadee, Newtownards, Comber, Carryduff, Yates Corner.
   Reference: Pollock and Wilson 1961 [O.S. 6: G.S. 29, 37]. About 80 km.

2. *Co. Tyrone*

   Belfast, Dungannon, Pomeroy, Carrickmore, minor roads to Tanderagee - Granagh - Coolnagreena - Drumnakilly - Glenscollip, Omagh, Seskinore, Clogher.
   References: Fearnsides and others 1907; Hartley 1933. [O.S. 5: G.S. 26, 34]. About 240 km.

C. Carboniferous Rocks

1. *Ballycastle area*

   Belfast, Ballymena, Ballycastle. Coastal section from Ballycastle to Murlough Bay.

Reference: Wilson and Robbie 1966. [O.S. 1: G. S. 8]. About 190 km.

2. *West Ulster*

Belfast, Coalisland, Dungannon, Enniskillen, Derrygonnelly, Garrison, Belcoo, Letterbreen, Florencecourt, Marble Arch, Lisnaskea, Augher, Aughnacloy, Benburb, Armagh.

References: Fowler and Robbie 1961; George 1953; Oswald 1935; Padget 1951–2–3. [O.S. 4, 5, 7. G.S. 35, 44, 45, 46, 56, 57]. About 450 km.

D. Mesozoic Rocks

Belfast, Collin Glen, Whitehead, Island Magee, Larne, Ballycastle, Ballintoy, Portrush, Coleraine, Limavady, minor roads to Keady Mountain, Dungiven, Maghera, Dungannon, Moira, minor roads to Lisburn.

References: Hancock 1961; Manning and others 1970; Reid 1962–4; Wilson and Robbie 1966. [O.S. 1, 2, 3, 5. 6. G.S. 7, 8, 12, 14, 18, 20, 21, 27, 29, 36]. About 380 km.

E. Igneous Rocks

1. *Co. Antrim*

Belfast, Ballynure, Carneal, Glenoe, Larne, Ballygally, Cushendall, minor road to Tievebulliagh, Ballyvoy, Fair Head, Carrickarade, Ballintoy Harbour, Giant's Causeway, Portrush, Ballymena, Slemish.

References: Agrell and Langley 1958; Eyles 1952; Harris 1937; Patterson 1955a, b, 1957, 1963; Preston 1963; Tomkeieff 1935, 1940, 1947, 1948, 1953; Wilson and Robbie 1966 [O.S. 1, 3: G.S. 7, 8, 14, 20, 21, 29]. About 300 km.

2. *Co. Down*

Belfast, Newcastle, Kilkeel, Hilltown, Rostrevor, Newry, Camlough, Lislea, Forkhill, Drumintee (forestry road), Slieve Gullion, Newry, Goraghwood Station.

References: Bailey 1956, 1959; Brown 1956; Elwell 1958; Emeleus 1955, 1962; Preston 1962; Reynolds 1943· 1957; Richey 1927, 1932; Tomkeieff 1935. [O.S. 6, 8, 9: G.S. 37, 60, 61, 70, 71]. About 210 km.

F. Quaternary feature

Belfast, Glenarm, Loughaveema, Carey River, Ballyvoy, Fair Head, Breen, Armoy, Coleraine, Binevenagh, Draperstown, Glenelly Valley, Gortin, Carrickmore, Pomeroy, Dungannon.

References: Charlesworth 1924, 1939, 1963; Stephens 1957, 1966; Wilson and Robbie 1966. [O.S. 1, 2, 3, 4, 5. G.S. 6, 7, 8, 12, 18, 25, 26, 34]. About 400 km.

19. Index

Publications of the Geological Survey of Northern Ireland

Memoirs

The Composition and Origin of the Antrim Laterites and Bauxites—62½p. The Geology of the Country round Dungannon—£1.50. The Geology of the Country round Ballycastle—£2.00. The Geology of Belfast and the Lagan Valley—£2.00. Regional Geology of Northern Ireland—50p.

Obtainable from Government Bookshops in Belfast, London, Edinburgh, Cardiff, Manchester, Birmingham and Bristol or any bookseller. Post orders to Government Bookshop, Chichester Street, Belfast, BT1 4JX.

Geological Maps

These maps are supplied by post from the Ordnance Survey of Northern Ireland, Ladas Drive, Belfast, BT6 9FJ, but are not available from Ordnance Survey agents. They can be bought at the Institute of Geological Sciences Bookstall, Exhibition Road, South Kensington, London, SW7.

Quarter-inch to one mile (1/253 440) Maps of Northern Ireland

Sheet 2 (1952) Parts of Antrim, Londonderry and Donegal, with southern part of Kintyre. Price 70p.

Sheet 5 Parts of Tyrone, Antrim, Armagh, Monaghan and Down. Reprinted. Price 70p.

Mineral Deposits Map (Provisional), 1970. Dyeline prints 40p.

One-inch to one mile (1/63 360) Maps

(First colour-printed editions with Irish Grid). Price 60p each sheet. Published in Solid and Drift editions.

Sheet 7 Giant's Causeway, 1968
Sheet 29 Carrickfergus, 1968
Sheet 36 Belfast, 1966
Sheet 8 Ballycastle, 1963
Sheet 35 Dungannon, 1961

Black and white line copies of the one-inch maps of the Geological Survey of Ireland covering Northern Ireland, except those replaced by colour-printed editions, are available from Ordnance Survey of Northern Ireland, Ladas Drive, Belfast, BT6 9FJ.

Three inches to one mile (1/21 120) Map

Special Engineering Geology Sheet, Belfast. 1971. £1.50.

Six inches to one mile (1/10 560) Maps

These sheets may be obtained by post from the Ordnance Survey of Northern Ireland. They are not available from I.G.S. Bookstall or Ordnance Survey Agents. (Uncoloured) Price 60p each sheet.

Tyrone 48 SE: 47 NW, SW; 54 NE
Antrim 5 SW, SE; 9 NW.

Quarter-inch to one mile (1/253 440) Geophysical Maps of Northern Ireland

Gravity Anomaly Map, 1967. Price 75p.
Aeromagnetic Map, 1970. Price 75p.

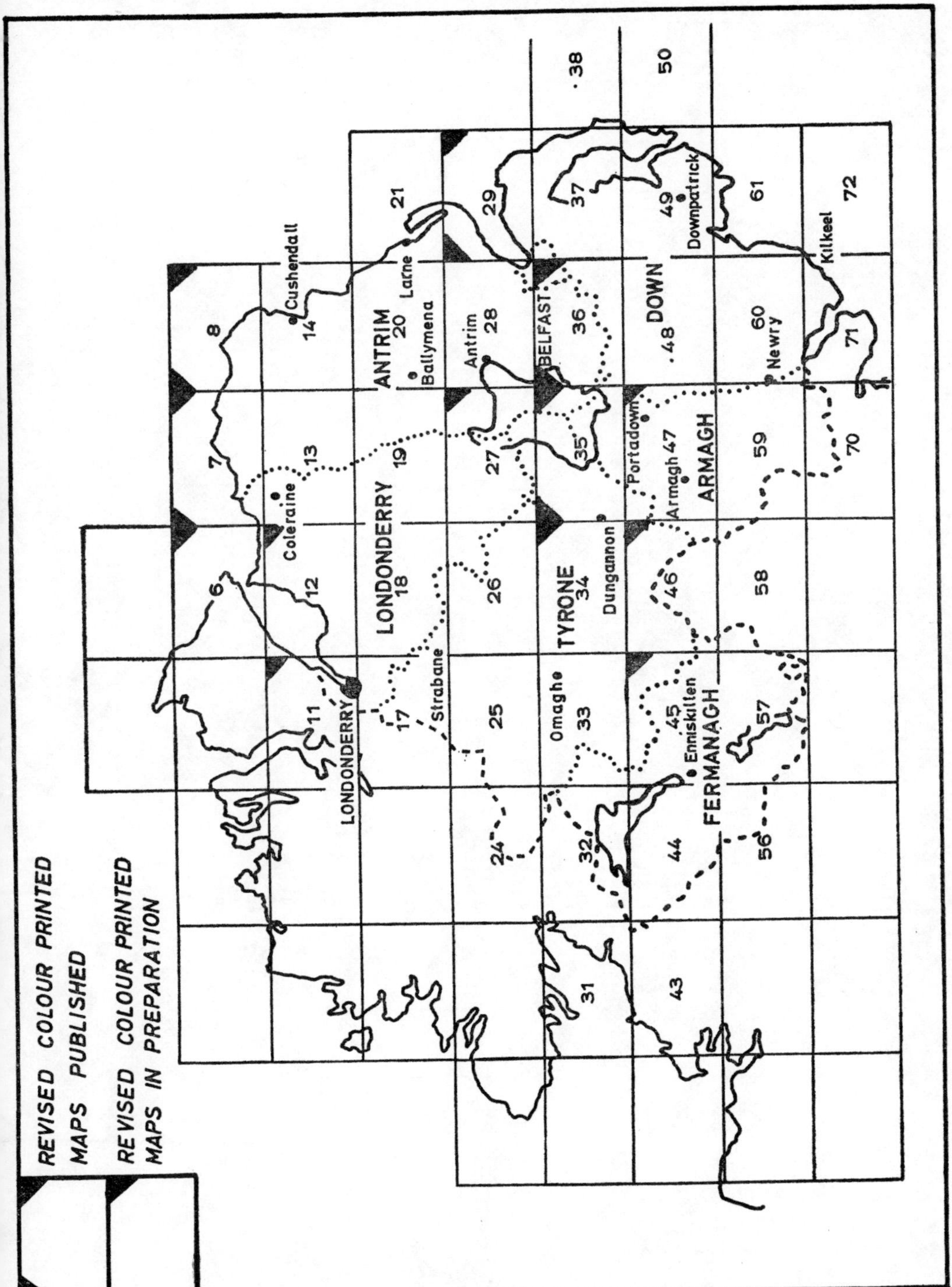
REVISED COLOUR PRINTED MAPS PUBLISHED
REVISED COLOUR PRINTED MAPS IN PREPARATION
LONDONDERRY
ANTRIM
TYRONE
DOWN
ARMAGH
FERMANAGH
Cushendall
Larne
Ballymena
Antrim
BELFAST
Coleraine
Strabane
Omaghe
Dungannon
Portadown
Armagh
Enniskillen
Newry
Kilkeel
Downpatrick
6
7
8
11
12
13
14
17
18
19
20
21
24
25
26
27
28
29
31
32
33
34
35
36
37
38
43
44
45
46
47
48
49
50
56
57
58
59
60
61
70
71
72